Colour Atlas of Glacial Phenomena

Michael J. Hambrey and Jürg C. Alean

CRC Press
Taylor & Francis Group
Boca Raton London New York

CRC Press is an imprint of the
Taylor & Francis Group, an **informa** business

CRC Press
Taylor & Francis Group
6000 Broken Sound Parkway NW, Suite 300
Boca Raton, FL 33487-2742

First issued in paperback 2017

© 2017 by Taylor & Francis Group, LLC
CRC Press is an imprint of Taylor & Francis Group, an Informa business

No claim to original U.S. Government works

ISBN-13: 978-1-4822-3440-4 (hbk)
ISBN-13: 978-1-138-48160-2 (pbk)

Dedication

Wilfred Theaksone, PhD supervisor at Manchester University, United Kingdom, and Geoffrey Milnes, post-doctoral mentor at ETH Zürich, Switzerland, for inspiring appreciation of glaciers and their landscapes; and to past and present staff, post-doctoral fellows and graduate students of the Centre for Glaciology at Aberystwyth University, Wales, for many years of fruitful collaboration on glaciological fieldwork.

—Michael J. Hambrey

Miles Ecclestone of Trent University, Canada, who is base camp manager on Axel Heiberg Island, and a good friend.

—Jürg Alean

FRONTISPIECE.   Tokositna Glacier descending southwards from the Alaska Range displays a multitude of lateral and medial moraines. North America's highest mountain, Denali (formerly Mount McKinley), rises to 6190 metres (20,310 feet) in the background. Photograph taken in July 2011.

Contents

Foreword

Glaciers and ice sheets are among the world's most beautiful and fascinating natural wonders. Whereas today glacier ice covers approximately a tenth of Earth's land surface, during the ice ages of the last few million years, as much as a third of the land surface was covered by glacier ice.

Modern glaciers and ice sheets possess many interesting features that reflect their dynamics, their interaction with landscape, rock, sediment and water, and they also inform us about climate change. The legacy of ice masses is reflected in the presence of a variety of landforms and sediments, many of which are also recorded in the rock record. The value of glaciers to society is immense. Glacial meltwater provides hydro-electric power, irrigates areas of low or seasonal rainfall; glacial sediments yield mineral-rich soil and provide abundant sand and gravel resources; and glaciers have shaped erosional landscapes that support tourism. Glaciers also threaten human life and property in mountain regions, and most important of all they influence global climate and sea level.

Several hundred phenomena associated with glaciers, both past and present, are described in this Atlas. With over 600 photographs, mainly from the authors' own collections accumulated over a period of forty years, glaciers and landforms in many parts of the world are illustrated: amongst the areas covered are the High-Arctic, the Western Cordillera of North America, the Andes of South America, the Alps of Europe, the British Isles, the Himalaya, New Zealand and Antarctica. In addition, certain glacier-related concepts are illustrated with diagrams. All these topics are preceded by an explanation of the various aspects of glacier science, thus setting the scene for the A–Z of glacial phenomena that forms the core of the book.

Preface

When we started our careers in glaciology over four decades ago, glaciers rarely figured in the public's imagination, except when people were directly affected by catastrophic events in mountain regions, or if they belonged to the mountaineering fraternity. Nowadays, with the realisation that glaciers are telling us about global warming and rising sea levels, glaciology has become a mainstream science. Barely a week goes by without reference in the media to disappearing glaciers and ice sheets.

Our own involvement in glaciology began as postgraduate students when we were tasked with investigating glaciers in Norway and Switzerland. We soon undertook fieldwork jointly in the Alps, and gradually began to develop a common interest in glacier photography that we have maintained to this day, despite living and working in different countries. We regard glaciers as among the most beautiful and fascinating elements of nature, responsible for creating some of the finest landscapes on Earth. As glaciers creep and slide down the mountains into the lowlands, they shape the landscape by scouring away bedrock, and carrying debris far from its source. Furthermore, glaciers provide water for hydroelectric power generation and irrigation, and many of the world's cities rely on them for a good proportion of their supplies. Glaciers also provide sediments that are useful in the construction industry, and on which mineral-rich soils can develop that are valuable for agriculture. On the other hand, glaciers can be hazardous; they have been responsible for disastrous ice avalanches and outburst floods that have caused heavy loss of life and damage to infrastructure. Today, it is more important than ever before to undertake research on glaciers. Glaciers almost everywhere in the world are receding at an accelerating rate. This is the result of the emission of man-made greenhouse gases into the atmosphere, the consequence of which is that global climate is warming up and sea levels are rising.

One of the joys of glaciology is that it is a multi-faceted subject, involving the disciplines of geology, geography, chemistry, physics, mathematics, biology and, increasingly, the social sciences. Consequently, there are many terms that students, professionals and the public may wish to understand. This *Colour Atlas of Glacial Phenomena* is a first attempt to provide a comprehensive summary of these terms. It stems from our website (www.glaciers-online.net), which we developed in our spare time for public outreach, and which has become widely used around the world. The website includes a 'Photo Glossary', and several colleagues suggested that we turn this into a book. Collecting material for this work proved challenging, with different terms or different definitions for the same feature, classifications of features that have evolved over time, terms that have been seemingly abandoned, and some that overlap other fields. We think it is quite likely that we have missed some terms, and defined others in a way that some colleagues may not find appropriate. We apologise if this is the case, and we would welcome any feedback from users of this volume.

The emphasis in this atlas is on features that can be seen in the field around the world, from the polar regions to the tropics. We cover terms related to glaciers and ice sheets, as well as the mountain and lowland landscapes that have been fashioned by ice. We include some conceptual terms, whilst simple diagrams supplement our photographs. We anticipate that the book will be useful to students and professionals alike, but we also hope that the wider public, with interests in the natural world, will enjoy looking at the multitude of features illustrated herein.

The atlas begins with a short introduction to glacial environments, which we have tried to pitch at a fairly 'gentle' level. However, we encourage the reader to use the atlas mainly as a companion to the various glaciological textbooks listed in the Introduction, where glaciological concepts are explained in more detail, supported by many examples and extensive reference lists. The bulk of the atlas, which follows the Introduction, is an A to Z compendium of all the glacier-related terms that we have discovered.

Michael J. Hambrey
Threlkeld, Cumbria, United Kingdom

Jürg Alean
Eglisau, Switzerland

Acknowledgements

FROM MICHAEL J. HAMBREY

Having recently retired to the English Lake District, where my interest in glacial environments was triggered as a teenager, I now look back on a glaciological career that began 45 years ago, bringing me into contact with some wonderful people, not only fellow university teachers, postgraduate students and post-doctoral scientists but also many interested non-specialists. I have spent months at a time in the field with many of them, often in tents in harsh conditions, sometimes wondering why we ever decided to 'do' glaciology, but always coming back to the realisation that glaciers are the most beautiful natural wonders on the planet. These folk are the rock on which I built my career at several institutions: the geography and geology departments at the University of Manchester; the Geological Institute at the Swiss Federal Institute of Technology (ETH) in Zürich; the Scott Polar Research Institute, Department of Earth Sciences and St Edmund's College at the University of Cambridge; the School of Biology and Earth Sciences at Liverpool John Moores University; and the Centre for Glaciology and Department of Geography and Earth Sciences at Aberystwyth University. I have also been privileged to spend extended periods at the Victoria University of Wellington and the University of Otago in New Zealand, and at the Alfred Wegener Institute in Bremerhaven, Germany, as a visiting fellow; and at the University of British Columbia, Canada, as a visiting professor.

It is impossible to thank everyone individually here, but they should be assured of my tremendous debt to them. However, there are some who stand out in the context of this book, through mentoring and providing unsurpassed opportunities to undertake fieldwork in remote parts of the world, and thereby obtaining many of the photographs used in this book. As the Dedication highlights, it was Wilfred Theakstone of Manchester University who first inspired my interest in glaciers and guided me through my PhD investigations on a glacier in northern Norway between 1970 and 1973. Then I thank Geoffrey Milnes, of the Swiss Federal Institute of Technology in Zürich, for the opportunity to work with him on alpine glaciers from 1974 to 1977, and for introducing me to a wide range of geological principles that could be applied to glaciers. In addition, there are the leaders of many research programmes and expeditions, who have provided me with the opportunities to undertake research in remote places. Having organised several expeditions myself, I can appreciate their unstinting and often unsung efforts to get such projects off the ground, from the initial writing a successful research proposal to organising the logistics; they deserve a special mention here (listed alphabetically by surname):

- **Peter Barrett** of the Victoria University of Wellington for the invitation to join two deep drilling projects (CIROS-1 and Cape Roberts) for determining palaeoclimate and glacial history in Antarctica in 1986 and 1997.
- **Garry Clarke** of the University of British Columbia for work on a surge-type glacier in the Yukon, Canada, in 2006.
- **Julian Dowdeswell** of the University of Cambridge for glaciological studies of surge-type glaciers in west-central Svalbard in 1996.
- **Werner Ehrmann**, **Dieter Fütterer** and **Gerhard Kuhn** of the Alfred Wegener Institute for Polar and Marine Research in Bremerhaven (Germany) for participation in a marine geological cruise of FS (RV) *Polarstern* to Antarctica in 1991.
- **Ian Fairchild** of the University of Birmingham for 'Snowball Earth' investigations in northeastern Svalbard in 2010 and Northeast Greenland in 2012.
- **Sean Fitzsimons** of the University of Otago, New Zealand, for glacial process investigations in the Dry Valleys, Antarctica, in 1999, and the McMurdo Ice Shelf in 2010.

- **Neil Glasser** of Aberystwyth University for glaciological investigations in Patagonia, Chile, in 1998, and the Antarctic Peninsula in 2012.
- The late **Brian Harland** of Cambridge University for participation in six geological expeditions to various parts of Svalbard between 1977 and 1983.
- **Barrie McKelvey** of University of New England, Armidale, Australia, for glacial geological fieldwork in the Prince Charles Mountains, East Antarctica, in 1994–1995.
- **Brian Moorman** for glaciological and glacial geological work on Bylot Island, Canadian Arctic.
- The late **Fritz Müller** of the Swiss Federal Institute of Technology, Zürich, for the opportunities to work on glaciers on Axel Heiberg Island in the Canadian Arctic in 1975.
- **Sean Richardson** (Swansea University) and **John Reynolds** (Reynolds International) for work on moraine-dammed lakes in the Cordillera Blanca, Peru, in 2002, and in the Everest region of Nepal in 2003.
- **Martin Sharp** then of Cambridge University (now University of Alberta, Edmonton) for the opportunity to work on an Alaskan glacier in 1986.
- **John Smellie** then of the British Antarctic Survey (now University of Leicester) for studies of glacier–volcano interaction on James Ross Island, Antarctica, in 2002 and 2006.
- **Peter Webb** of Ohio State University, United States, for glacial geological fieldwork in the central Transantarctic Mountains in 1995–1996.
- **Peter Worsley** of Reading University for glaciological work based at the Okstindan Research Station, northern Norway, in 1970–1973.

In addition, numerous other geologists are thanked for leading field excursions in which the author has participated, to various other glacier-influenced regions of the world. Numerous photographs that feature in this book were taken on these excursions. Many of the photographs could not have been taken but for the skills of the pilots of chartered helicopters and fixed-wing aircraft, and the boat crews of both large and small research vessels.

I am also indebted to various funding bodies and logistical organisations for making much of the research that lies behind the photographs possible: the UK Natural Environment Research Council, the US Antarctic Program, the New Zealand Antarctic Research Programme/Antarctica New Zealand, the Alfred Wegener Institute, the Australian National Research Expeditions, the British Antarctic Survey, the Leverhulme Trust, the European Centre for Arctic Environmental Research (*ARCFAC*), the Transantarctic Association, the Polar Continental Shelf Project (Canada), the Knowledge Transfer Partnership (United Kingdom), the Climate Change Consortium of Wales, the Swiss Centenarfond and the various universities that have hosted me.

FROM JÜRG ALEAN

My interest in glaciology was kindled at the Swiss Federal Institute of Technology in Zürich. As an undergraduate student of geography, I was given the unique opportunity, by the late Fritz Müller, to participate in two major expeditions to Axel Heiberg Island in the Canadian Arctic, where I undertook fieldwork on White and Baby Glaciers that led to my master's thesis. The solitude of this wonderful island, the fascinating beauty of its landscape, and not least its sparkling white glaciers are a lasting memory I treasure to this day, four decades later. Back in Switzerland, my PhD thesis, under the guidance of Wilfried Haeberli, attempted to improve our understanding of the formation and reach of ice avalanches.

My professional career, in the following years, was mainly outside the university context. However, as a teacher of geography at a Swiss grammar school, I was able to maintain many productive contacts with academia, and not least with my co-author, which led to the publication of several popular scientific books about glaciers. In 2008, I was most fortunate to be able to realise a career-long dream, which was to return to Axel Heiberg Island, specifically for glacier photography. Special thanks are due to Graham Cogley and Miles Ecclestone, both at Trent University, Peterborough, Canada, as well as Wayne Pollard at McGill University, Montreal, Canada, who made this possible. I am indebted to Atsumu Ohmura for catalysing these contacts. In 2009, I was also able to visit Svalbard as part of a project organised by Michael J. Hambrey, funded by the European Centre for Arctic Environmental Research (ARCFAC) and hosted by the German–French facility (AWIPEV) in Ny-Ålesund.

I also wish to thank Werner Hartmann, initiator of the educational web platform 'Swisseduc. ch' for his continued support in all matters involving computers and software. 'Swisseduc.ch' has been, and continues to be, an excellent platform for our voluntary outreach efforts in glaciology. Finally, I thank my dear wife, Pamela Alean, for her patience during the production of this book, for enduring flights in cold and open-windowed aeroplanes during glacier photo flights, and for general support during our global travels.

GENERAL

FOR PROVIDING PHOTOGRAPHS, WE WOULD LIKE TO THANK THE FOLLOWING COLLEAGUES:

Andrew Finlayson (British Geological Survey, Edinburgh, United Kingdom) for Figure R.27
Simon Cook (Manchester Metropolitan University, United Kingdom) for Figure A.4
Julian Dowdeswell (University of Cambridge, United Kingdom) for Figures D.13, F.49, M.5, G.49
Sam Doyle (Aberystwyth University) for Figures G.44a and b
Arwyn Edwards and Ottavia Cavalli (Aberystwyth University, United Kingdom) for Figure M.8a
David Evans (University of Durham, United Kingdom) for Figures C.32b, C.40
Edward Fleming (Cambridge Arctic Shelf Programme, United Kingdom) for Figure H.14
Marco Fulle (Osservatorio Astronomico, Trieste, Italy) for Figures J.2, S.60
Uli Herzog (Johannesburg, Germany) for Figure R.3
Bryn Hubbard (Aberystwyth University, United Kingdom) for Figure R.17b
Daniel Le Heron (Royal Holloway, University of London, United Kingdom) for Figure C.16a
Mauri Pelto (Nichols College, Dudley, Massachusetts, United States) for Figure I.50
Emrys Phillips (British Geological Survey, Edinburgh, United Kingdom) for Figure R.1
Helen Roberts (Aberystwyth University) for Figure L.11b
Margit Schwikowsky and Anja Eichler (Paul Scherrer Institute, Würenlingen, Switzerland) for Figure I.23
Marcus Weber (Bayerische Akademie der Wissenschaften, Glaciology, Munich, Germany) for Figure D.20
Heinz Zumbühl (University of Bern, Switzerland) for Figure L.8
For drafting of figures we thank Antony Smith of Aberystwyth University.

PERMISSION HAS BEEN GRANTED AS REQUIRED FROM THE FOLLOWING ORGANISATIONS:

Cambridge University Press, Antarctic Science, for Figure S.58
Haupt-Verlag, Bern, Switzerland, for Figures I.1 and I.2 (Introduction)
International Glaciological Society for Figure D.5
Natural Environment Research Council, United Kingdom, for Figures E.9b, F.26
Federal Office of Topography, Switzerland (Swisstopo), for Figure F.25

WE ALSO ACKNOWLEDGE THE FOLLOWING ADDITIONAL SOURCES:

The book *Wunderland Peru*, Hans Huber Verlag Bern, for Figure A.14
National Aeronautical and Space Administration (NASA) for Figures C.3, C.8, D.29, F.11, F.15b, G.35, I.18, I.34a, b, I.39, I.41, I.45, P.11, R.17a, S.55

NASA/JPL/University of Arizona for Figure G.26
Kluwer, Dordrecht, for Figure G.7
Geological Society of London for Figure M.11
US Geological Survey for Figure G.37
Natural Environment Research Council, United Kingdom, for Figures E.9, F.26.

About the Authors

Michael J. Hambrey (right) is emeritus professor of glaciology and former director of the Centre for Glaciology in the Department of Geography & Earth Sciences, at Aberystwyth University, Wales, UK. He was also founding director of the Climate Change Consortium of Wales. His research interest includes glacial geology and structural glaciology. He has published more than 200 scientific papers and several books, including a university-level textbook *Glacial Environments* (1994), as well as the popular books *Glaciers* (2nd edn, 2004) and *Gletscher der Welt* (2013) with Jürg Alean, and *Islands of the Arctic* (2002) with Julian Dowdeswell.

Mike has undertaken glaciological fieldwork in Norway, New Zealand, the Swiss Alps, the Andes, the Himalaya, the Canadian Arctic, Yukon, Alaska, Greenland, Svalbard and Antarctica. For his work in the polar regions, he has been awarded the 'Polar Medal' twice by Her Majesty the Queen (1989 and 2012). He has served on a number of UK national and international committees dealing with glacial and polar issues.

Jürg Alean (left) was formerly a teacher of geography at the Kantonsschule Zürcher Unterland in Bülach, Switzerland. He has undertaken extensive fieldwork in the Swiss Alps, the Canadian Arctic, Alaska and South America. His research has led to various scientific papers, in particular concerning dangerous glaciers and ice avalanches. He has also published many popular scientific articles and several books, for example, *Gletscher der Alpen* (2010) and *Gletscher der Welt* (2013). He is member of a team of volunteers maintaining 'SwissEduc.ch', a Web platform dedicated to providing teaching materials mainly for secondary education, and also hosting 'Glaciers-online. net', where both authors present a wide range of their photographic work on and around glaciers all over the world.

Mike and Jürg's book *Glaciers* (1st edn) earned the Earth Science Publishers (USA) Outstanding Publication Award in 1995.

Introduction to Glacial Phenomena

1 TYPES AND DISTRIBUTION OF GLACIERS

Glaciers are among the most fascinating natural phenomena on Planet Earth. They display a wide range of dynamic phenomena, gouge out the landscape, carry and deposit large volumes of sediment, and release large volumes of meltwater that also erode and deposit sediments. According to some estimates, approximately 10% of the world's land area is covered by **glacier ice**, and this reached around 30% at the peak of the last **ice age**, around 20,000 years ago. The legacy of **glaciation** is a vast array of landforms and sediments that can be used to reconstruct records of former climatic change. Furthermore, glaciers have not only produced some of the world's finest scenery but also benefitted human civilisation through provision of water, fertile soils, aggregates for construction purposes and opportunities for tourism. Glaciers have also been implicated in various natural disasters, but their biggest impact is yet to come: through the raising of global sea levels.

Glaciers range in size from small **niche glaciers** and **glacierets** that occupy small hollows and depressions, to the vast **ice sheet** covering Antarctica. However, most of us are probably most familiar with the more accessible **valley glaciers**, such as found in mountain regions today (Figure 1). Along with depositional and erosional products, all of these features are represented in this volume.

FIGURE 1   Snow-covered valley glacier descending from the peak of Dansketinde (2795 m), Stauning Alps, Northeast Greenland National Park (July 1985).

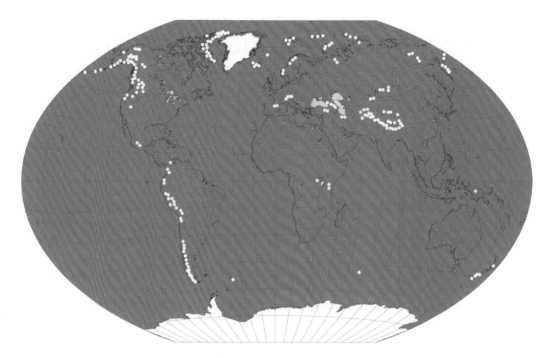

FIGURE 2   Global distribution of glaciers and ice sheets. (From Alean, J. and Hambrey, M.J., *Gletscher der Welt*, Haupt AG, Bern, Switzerland, 2013.)

The distribution of glaciers is almost global, ranging from the polar regions to the high mountains of the tropics (Figure 2). By far, the largest area of glacier ice is that in Antarctica (13.6 million km²), which covers almost the entire continent to an average depth of 1.8 km. Next follows the ice sheet over Greenland, occupying 1.7 million km². Together these ice sheets account for 99% of the world's ice. Extensive **highland icefields** and **ice caps** occur in central Asia, the Arctic, northwestern North America and Patagonia in South America. Small valley glaciers, Bersærkerbræ, representing only a small fraction of the total ice cover, are scattered across the mountain ranges of the rest of the world.

Apart from glaciers themselves, considerable pioneering work on glacial landforms and processes has been undertaken in much larger areas formerly covered by glaciers, and many of the terms used are a legacy of researchers from these areas. Only a small proportion of the terms explained in this atlas are mentioned in this Introduction. As here, they are indicated in bold when first used in each section, for ease of cross-referencing. However, tables are provided for all the terms relevant to each theme. Inevitably, many terms belong to more than one category. The terms in this atlas relevant to the different types of glaciers and their distribution are listed in Table 1.

2   GLACIERS IN EARTH'S HISTORY

This atlas illustrates many features related to past glaciations (Table 2). Their interpretation is largely based on our knowledge of modern glacial processes. As is widely known, Earth's climate has been constantly changing, as a consequence of which our planet has experienced several continental-scale glacial periods, as well as periods without ice. Indeed, the geological record, as a whole, shows that an absence of ice has been the norm, and that the conditions we now live under are atypical.

The earliest known rocks on Earth, in Australia, date from 4.4 billion years ago, and over the entire span of geological time there have been five major and several minor glacial eras. The main ones occurred in the early and late Proterozoic Eon, around the Ordovician–Silurian boundary, spanning most of the Permian and Carboniferous periods, and in the second half of

TABLE 1
Terms Relevant to Glacier Types and Distribution

Alpine glacier	Ice cliff	Piedmont glacier
Avalanche-fed glacier	Ice divide	Piedmont moraine
Benchmark glacier	Ice dome	Plateau ice cap
Calving glacier	Ice fringe	Polar glacier
Cirque glacier	Ice front	Polythermal glacier
Cold glacier	Icehouse state	Reconstructed glacier
Crater glacier	Ice island	Reference glacier
Dead ice	Ice piedmont	Regenerated glacier
Dry-based glacier	Ice rise	Remnant glacier
Expanded foot glacier	Ice rumple	Subpolar glacier
Freshwater glacier	Ice sheet	Surge-type glacier
Glacier	Ice shelf	Surging glacier
Glacier complex	Ice stream	Temperate glacier
Glacier-like forms (Mars)	Ice tongue	Terminus
Glacier–volcano interaction	Ice wall	Tidewater glacier
Glacieret	Index glacier	Toe
Glacierised/glacierized	Inland ice	Transection glaciers
Hanging glacier	Mountain-foot glacier	Valley glacier
Highland icefield	Mountain glacier	Warm glacier
Ice	Niche glacier	Warm-based glacier
Ice cap	Outlet glacier	Wet-based glacier

TABLE 2
Terms Relevant to Glaciers in Earth's History

Core-logging	Holocene	Pleistocene
Eccentricity	Ice age	Precession of the equinoxes
Fluctuations	Icehouse state	Quaternary Period
Glacial period/glaciation	Interglacial period	Sequence stratigraphy
Glacial sedimentary logging	Isostasy	Snowball Earth
Glaciated	Little Ice Age	Stadial (or stade)
Glaciated landscape	Obliquity of the ecliptic	Till
Glacio-eustacy	Palaeoglaciology	Tillite
Glacioisostacy	Paraglacial processes	Younger Dryas Stadial

the Cenozoic Era. Of these, the last embraces the well-known Quaternary **ice ages** when ice sheets spread over much of the Northern Hemisphere. In the older geological record, the evidence for glaciation is in the form of **tillites** – sedimentary rocks demonstrably of glacial origin – as well as related deposits and glacial erosional surfaces.

The earliest evidence for glaciation is confined to South Africa, and dates from about 3000 Ma. The first continent-wide glaciation spanned a long interval of time in the early Proterozoic Eon, from 2500 to 2000 Ma, and affected North America, South Africa (again) and Western Australia. Of these, the best studied and most impressive are those exposed at Lake Huron in Canada, where the most recent glaciations of the Quaternary Period have created freshly exposed surfaces of this ancient glaciation.

Arguably the most intense and widespread phase of glaciation occurred during the late Proterozoic (Neoproterozoic) Eon, spanning the interval of at least 750–600 Ma. Tillites occur on every continent, and many superb geological sections have been documented. Although there are uncertainties of dating and correlation, the glacial rocks seem to group mainly, but not exclusively, around the time intervals 720–700 Ma and 660–635 Ma, and they are known, respectively, as the Sturtian and Marinoan glaciations, after sites in Australia. The global extent of these events has led to the development of the **Snowball Earth** theory, which envisages a totally ice-covered Earth and lockdown of the hydrological cycle. Palaeomagnetic evidence is used to determine global extent of tillites, and the Snowball Earth theory explains why they occur in tropical regions. However, the theory is not universally accepted, and many geologists argue for more localised ice cover related to tectonic processes.

The next glacial period of wide extent occurred in latest Ordovician and earliest Silurian time (440–450 Ma). It affected Africa, the Arabian Peninsula, Europe, and North and South America. Knowing the approximate arrangement of these regions within the precursor to the Gondwana supercontinent, this can be explained as a polar glaciation. However, good palaeontological control suggests that extensive ice may have existed for only a few million years.

More extensive and longer lasting was the next glacial period in the Carboniferous and Permian periods (350–250 Ma). Over this near-100-million-year period, ice sheets waxed and waned over the Gondwana supercontinent, which embraced Africa, Antarctica, southern Asia, Australia and South America. This, too, was a South Polar glaciation, but timing of glaciation at any one place may have been influenced by proximity to the South Pole, with the earliest ice sheets growing over South America, and the latest in Australia. This Permo-Carboniferous glaciation is the best known of the ancient glaciations, largely because of the remarkable preservation of glacial erosional features (Figure 3a).

FIGURE 3 Evidence of former large-scale glaciations. (a) An exhumed glacially striated pavement near Douglas in the great Karoo region of South Africa. This was produced during the Permo-Carboniferous ice age that affected the Southern Hemisphere continents when they were united in the supercontinent of Gondwana, then located in the South Polar region.   *(Continued)*

(a)

(b)

FIGURE 3 (Continued) (b) One of the many thick Quaternary sequences dating from the last Quaternary ice age in Britain, the Devensian of around 20,000 years ago. These crumbling cliffs at Robin Hood's Bay, Yorkshire, comprise glacial deposits (till), sand and gravel, which are vulnerable to erosion by the sea.

This glaciation is implicated in the formation of coal and limestone, since changing sea levels (as ice sheets waxed and waned) resulted in the formation of extensive shallow seas and organic-rich swamps. The exploitation of these resources in Europe triggered the Industrial Revolution.

A long phase of warm climates prevailed after the end of this glaciation, and, from 235–34 Ma, spanning the whole Mesozoic and the early Cenozoic eras, there are only a few indications of sea ice, and certainly no widespread glacier ice cover. Global cooling from the start of the Cenozoic Era (65 Ma) did not trigger the first signs of glaciation until the first Antarctic Ice Sheet developed around 34 Ma. The evidence for the onset and progression of this glaciation was obtained from long sediment cores drilled on the continental shelf of Antarctica. These cores indicate that the first ice sheet was temperate, but then subsequently, cooling eventually yielded the cold ice sheet that we have today. In contrast, the first ice sheet in the Northern Hemisphere, over southern Greenland, did not grow until much later, probably around 10 Ma, and it was not until around 3.5 Ma that this grew to cover the whole island, as today. After this, other ice sheets expanded into temperate latitudes, in turn over northern Eurasia around 2.74 Ma and North America at 2.54 Ma. These timings have been determined from records of **iceberg-rafting** found in deep-sea cores, so they do not necessarily preclude ice sheets in the interior of these continents.

The Quaternary Period has traditionally been regarded as synonymous with the recent 'ice ages' or 'glaciations' (also referred to as 'glacials'), but it was only in 2009 that its status and start date was formally defined – at 2.58 Ma. Since then, Earth's climate has fluctuated through glacial and interglacial cycles but, as mentioned in the preceding text, these were already occurring in Antarctica and Greenland well before this date. Historically, only four ice ages were recognised in the terrestrial records of North America and Europe, but these records are incomplete. However, extraction of drill-cores from the deep ocean and analysis of their **oxygen isotope records** (which indirectly indicate ice volume and temperature) has yielded at least 50 cold/temperate cycles, of which only a few are documented on land. By far the best known Quaternary glaciation is the most recent, since successive glaciations have tended to destroy much of the evidence of earlier ones on land. Different names for the last glaciation apply in different regions, for example, Wisconsinan in North America, Weichselian in northwestern Europe, Würm in the Alps and Devensian in Britain. At this time, ice sheets covered much of the Northern Hemisphere, but earlier events were even more extensive. This glaciation peaked around 18,000–20,000 years ago, a time referred to as the **Last Glacial Maximum** (Figure 3b). Since then, the ice sheets have waned with interruptions, and the remaining ice sheets are now confined to the polar regions. However, as noted in Section 1, many mountain glaciers and ice caps survive.

3   GLACIERS AND CLIMATE

Glaciers are generally regarded as one of the best indicators of climate change. They are not subject to the vagaries of year-on-year variations, but respond mainly to decadal scale changes in climate. In a warming climate, ice masses also contribute one of the two main sources of sea level rise, the other being thermal expansion of the oceans.

The traditional method of evaluating the response of glaciers to climate change has been through **mass balance** studies, and there are many terms relevant to this approach (Table 3). These studies entail measuring the inputs of snow and the outputs of melting ice to determine changes in glacier mass. Globally, relatively few glaciers have been monitored in this way (Figure 4), but some records

TABLE 3
Terms Relevant to Climate Change

Ablation	Glacier (or glacial) advance	Reference glacier
Accumulation	Glacier–climate interaction	Response time
Accumulation area (zone)	Glacier fluctuations	Retreat
Accumulation-area ratio	Glacier health	Sequence stratigraphy
Advance	Glacier recession (retreat)	Shelf ice
Albedo	GLOF	Snout
Balanced budget	Health of a glacier	Snowline
Benchmark glacier	Holocene	Snow stratigraphy
Cold glacier	Ice age	Sublimation
Cryosphere	Ice core	Summer surface
Cumulative mass balance	Icehouse state	Superimposed ice
Dead ice	Index glacier	Temperate glacier
Diurnal fluctuation/variation	Interglacial period	Terminus
Dry-snow zone	Little Ice Age	Thermal regime
Dust	Mass balance	Toe
Ecological succession	Obliquity of the ecliptic	Tongue
Energy balance	Oxygen isotope analysis	Transient snowline
Equilibrium line/zone	Palaeoglaciology	Trimline
Firn	Pleistocene	Water equivalent
Firn line	Polythermal glacier	Wet snow zone
Fluctuations	Precession of the equinoxes	Wind accumulation
Glacial lake outburst floods	Quaternary Period	Winter advance
Glacial period/glaciation	Reaction time	Younger Dryas Stadial
Glacier	Recession	

FIGURE 4 White Glacier on Axel Heiberg Island in the Canadian Arctic. This glacier has one of the longest mass-balance records in the world, dating from 1960. Such records are crucial in determining the response of glaciers to climate change (July 2008).

go back to the 1940s. Mass balance studies are labour-intensive, and the fieldwork is both challenging and expensive. Other monitoring studies involve simply measuring the position of a glacier snout annually, usually in late summer. Some of these records go back into the mid-nineteenth century, notably those documented by the Commission of Glaciology of the Swiss Academy of Natural Science, with data deposited in the World Glacier Monitoring Service, and these provide an excellent summary of long-term trends. When combined with mapping of ice limits using moraines and applying numerical modelling approaches, values of former ice mass can be determined. Mapping of moraines and dating them using radiocarbon and cosmogenic nuclide dating techniques allows **glacier fluctuations** to be determined over several centuries or millennia, or over a few million years in the case of the latter technique. Aerial photographs, satellite images and even photographs taken from the International Space Station are the principal sources needed to undertake the mapping (Figure 5).

Although terrestrial glaciers show the clearest response to climate change, one group of glaciers, namely those that are **surge-type**, behaves in a way that is unrelated to climate. Rather, they reflect short periods of dynamic instability induced by changes in thermal regime or basal hydrology. For example, several glaciers in Svalbard have surged in recent years, even though most glaciers are receding strongly, and some have even lost their **accumulation areas**. Similarly, **tidewater glaciers** do not respond to climate change in a clear manner. For example, in a warming climate, they can remain quite stable for many years if anchored to the sea bed, but then undergo catastrophic **recession** into deeper water once a critical limit of thinning has been reached (Figure 6).

Field-based mass balance methods cannot be applied to large valley glaciers for logistical reasons, let alone to large ice sheets. However, recent decades have seen the emergence of an array of remote sensing techniques applicable to monitoring the response of large glaciers and ice sheets to climate change. Rapidly developing technology and deployment of new satellites allow surface elevations of ice sheets to be determined, for example, through laser altimetry. However, with small snowfalls in the centre of ice sheets, the accuracy of these methods remains a serious limitation. Alternative methods

FIGURE 5   The 350-km-long Campo de hielo Patagónico Sur (Southern Patagonia Icefield), covering 13,000 km², is the largest temperate ice mass in the Southern Hemisphere. During the last ice age, the icefield was considerably larger and eroded out the fjords in Chile (upper) and lakes in Argentina (lower). This photograph was taken by a crew member on the International Space Station on 13 February 2014 (NASA image no. ISS038-E-47324).

FIGURE 6   Glacier recession is commonly most rapid in tidewater glaciers such as Conwaybreen, northwestern Spitsbergen, Svalbard. In such cases, the climatic signal of recession is complicated by bathymetry and oceanographic changes (July 2009).

use satellites that make gravimetric measurements, using the principle that land rebounds as ice cover is reduced (**glacio-isostatic adjustment**), and allowing mass changes of ice sheets to be determined.

Longer-term climate records are needed to make reliable future predictions about glacier health and contribution to sea level. The most comprehensive records of climate change during the Quaternary Period are linked to fluctuations of glaciers and ice sheets. *Palaeoclimatology* is the study of past climates, and it uses data obtained primarily from deep-sea sediment **core-logging** (spanning millions of years) and ice cores from the Greenland and Antarctic ice sheets (the longest spanning 800,000 years). The oxygen isotope records obtained from these cores, using marine organisms (foraminifera) and trapped atmospheric gases, respectively, correlate with changes in temperature and ice volume. Additional palaeoclimatic information comes from long sediment cores drilled on the continental shelf of Antarctica (providing a direct record of glacier fluctuations over tens of millions of years), ice cores from ice caps (for records of climate and pollution on time scales of a few thousand years), lake sediment cores and pollen records (especially for interglacial and postglacial climates).

The most important contribution from these studies has been the recognition that, during the Quaternary Period, ice volumes and global temperatures were intimately linked with carbon dioxide concentrations in the atmosphere. During the past 800,000 years, until the Industrial Revolution, carbon dioxide concentrations have varied between 180 and 300 ppm, and these match glacial/interglacial cycles. These cycles are driven by variations in solar radiation reaching Earth's surface as a result of the interaction of three astronomical elements, namely eccentricity of the orbit, tilt of the ecliptic and precession of the equinoxes; these cycles were originally defined theoretically by the Serbian geophysicist Milanković in the 1920s. Now, as a result of adding this gas to the atmosphere from fossil fuel use and from destruction of forests, the concentration has reached 400 ppm (in 2015), and is rising rapidly, along with global temperatures. The human-induced output of another more potent, but shorter-lived greenhouse gas, methane, is also contributing to global warming, with thawing of permafrost exacerbating the problem. As a consequence, mountain glaciers and ice caps, with only a few exceptions, are receding and thinning rapidly, and perhaps more seriously losing their accumulation areas. Indeed, climate scientists and glaciologists predict that most of the smaller ice masses will disappear by the end of the century. Of the ice sheets, those over Greenland and West Antarctica are losing mass at an accelerating rate, potentially adding several metres to sea-level rise. The state of the East Antarctic Ice Sheet is less certain: some analyses suggest it is thinning, and others that it is thickening as a result of increased snowfall. What is not in doubt is that sea levels are rising at an accelerating rate, indicating both increasing net ice-mass loss and thermal expansion of the oceans. Arguably, stemming these changes, and adapting to them, are humanity's biggest challenges.

4   GLACIER DYNAMICS AND DEBRIS ENTRAINMENT

A fascinating array of features, illustrated in this atlas (Table 4), arises from the fact that glaciers are, or have been, highly dynamic. 'Healthy' glaciers flow at rates typically of tens or hundreds of metres a year, and in exceptional cases at several kilometres per year. The fastest glaciers are **outlet glaciers** from the Greenland Ice Sheet that terminate in the sea. There is also a type of glacier that switches from slow to fast modes on time scales varying from a decade to a century; these **surge-type glaciers** have speeds that may reach several hundred metres per year. There is, however, an increasing proportion of glaciers that are stagnant, wasting away *in situ* as rapid thinning and recession take hold.

Glacier ice forms as a result of the build-up of layers of snow, which is then subject to compaction and metamorphism, through an intermediate state called **firn**. The ice initially is layered or **stratified**, as in a sedimentary rock, but is soon modified as the ice flows under the influence of gravity. Glaciers flow by three main mechanisms: **internal deformation** or **creep**, **sliding** over a rigid bed as a result of lubrication by water, or by movement over a soft **deformable bed** of sediment. Internal deformation produces a wide range of structures. These include the products of brittle deformation, such as **crevasses** and **faults** (Figure 7), and those that result from ductile deformation, including **folds** and **foliation**. All these features, and many others, are described and illustrated in this atlas.

TABLE 4
Terms Relevant to Glacier Dynamics

Ablation area crevasses	Debris apron	Glacier sole
Abrasion	Debris entrainment	Glacier terminus
Accreted/accretion ice	Debris layering	Glacier tongue
Adfreezing	Debris-mantled (Debris-covered) glacier	Glacier–volcano interaction
Advance	Debris septa	Glacierised/glacierized
Arcuate crevasses	Deformable bed	Glaciohydraulic supercooling
Arcuate (transverse) foliation	Diagenesis	Glen's flow law
Ash-layering	Dirt cone (see debris cone)	Grounding-line/zone
Avalanche	Diurnal fluctuation/variation	Hot-water drilling
Band ogives	Downwasting	Hydrofracturing
Basal accretion	Dry-based glacier	Ice
Basal crevasse	Dry calving	Iceberg
Basal debris/sediment	Dump moraine	Iceberg armada
Basal glide	Dynamic thinning	Iceberg crevasses
Basal hydrology	En echelon crevasses	Iceberg rafting or ice-rafting
Basal ice layer	Englacial debris	Ice blister
Basal melting	Equilibrium line/zone	Ice breccia
Basal plane	Extending flow	Ice cliff
Basal shear stress	Fabric	Ice core
Basal sliding	Facies (basal ice)	Ice crystals
Bedrock step	Facies (glacier ice)	Ice crystal fabric
Bed roughness	Fast glacier flow	Ice crystallography
Bergschrund	Fault (glacier ice)	Ice divide
Bergy bit	Firn	Ice facies
Blue ice	Floating tongue	Icefall
Blue ice area	Flow bands	Ice front
Boudin/boudinage	Flow law	Ice island
Brash ice	Flowlines	Ice mélange
Brittle deformation	Flow stripes	Ice-rafted debris
Bulldozing	Flow unit	Ice rumple
Calving	Fluctuations	Ice-shelf collapse
Calving glacier	Fold	Ice stagnation
Cauldron collapse	Foliation	Ice tongue
Cavitation	Forbes bands	Ice vein
Chevron crevasses	Fractures (ice)	Inter-crevasse blocks
Chevron folding	Frictional heat	Internal deformation
Cold glacier	Geothermal heat	Isoclinal folding
Cold ice	Glacial erosion	Kinematic wave
Compressive (or compressing) flow	Glacier	Longitudinal crevasses
Concentric crevasses	Glacier (or glacial) advance	Longitudinal foliation
Congelation ice	Glacier dynamics	Looped moraine
Creep	Glacier flow/motion	Low-angle crevasses
Crevasse	Glacier fluctuations	Marginal crevasses
Crevasse trace	Glacier front	Medial moraine
Crushing	Glacier hydrology	Parasitic folds
Crystal growth	Glacier ice	Pinning point
Cumulative strain	Glacier margin	Plasticity
Dead ice	Glacier recession (retreat)	Plucking

(Continued)

TABLE 4 *(Continued)*

Terms Relevant to Glacier Dynamics

Polish	Shear zone	Surge loops
Polythermal glacier	Sheath fold	Surge moraine
Pressure melting point	Similar fold	Surge-type glacier
Pressure ridge	Simple shear	Surging glacier
Primary structures	Sliding	Syncline
Pure shear	Slow surge	Tabular berg
Push moraine	Snout	Teardrop moraines
Quarrying	Snow bridge	Temperate glacier
Receiving area	Sole	Terminus
Recession	Splaying crevasses	Thermal regime
Recrystallisation	Sticky spot	Thrust (ice)
Recumbent fold	Strain	Toe
Refolded fold	Strain ellipse	Tongue
Regelation	Strain net	Total strain
Reservoir area	Stratification	Transverse crevasses
Response time	Streaklines	Transverse foliation
Retreat	Stress	Unconformity
Run-out distance	Subglacial volcanism	Velocity
Secondary structures	Subglacial water pressure	Viscosity
Sedimentary stratification	Subpolar glacier	Wave ogives
Sérac	Summer surface	Yield stress
Shear	Surge	Zone of suspension
Shear margin	Surge front	Zone of traction

FIGURE 7 Extensive crevassing in the upper reaches of Kahiltna Glacier, Denali National Park, Alaska, indicates the fast-flowing nature of this large valley glacier (July 2011).

FIGURE 8   Basal debris being entrained from the base of Conwaybreen, northwestern Spitsbergen, Svalbard (July 2009).

The dynamics of glaciers vary according to the climatic regime in which they are located. Glacier-favourable climates range from temperate-maritime to cold-arid polar. The thermal characteristics of glacier have a profound influence on their ability to slide on their bed. In temperate regions such as the Western Cordillera of North America, or the European Alps and the Southern Alps of New Zealand, glaciers are at the pressure-melting point throughout (except for a temporary winter cold wave in the surface layers). They are referred to as **temperate** or **warm glaciers**, and most of the ice movement is achieved by **basal sliding**. At the other extreme are the cold polar ice masses, notably in Antarctica, where most of the ice is below the pressure melting point, and flow is predominantly by internal deformation. These **cold glaciers** are frozen to the bed, although it should be noted that the thickest ice may be melting at the base as a result of geothermal heating. A glacier with intermediate characteristics, that is, part-cold and part-warm, is referred to as **polythermal**, and these are especially common in the high Arctic and at high altitude, such as in the Himalaya.

Glaciers are notable for their ability to erode, entrain and transport debris. Glacial erosion is achieved by the impact of entrained debris at the base of the glacier. A wide range of erosional landforms is the result (Section 5). Temperate glaciers, and to some extent polythermal glaciers, are effective erosional agents. In contrast, cold glaciers tend to protect the landscape, although in some circumstances stresses induced by internal deformation, may propagate into the sediment or bedrock below and also cause significant erosion.

Debris is entrained in the base of a glacier by the process of freeze-on, especially where pressure changes lead to melting and refreezing. This process gives rise to a **basal debris layer** that can reach several metres in thickness. Further uplift into a glacier is achieved by **folding** and **thrusting** (Figure 8). Debris also collects on the surface of a glacier from rockfall, producing **lateral moraines**, and where tributary glaciers join, **medial moraines** (Figure 9). This debris also may enter an **englacial** position, by falling into crevasses and **meltwater channels**. The way in which a glacier carries debris is reflected in the resulting landforms (Section 6).

FIGURE 9 Supraglacial debris, derived from rockfall, on the surface of Grosser Aletschgletscher, Switzerland, arranged in a series of neat medial moraines (October 2006).

5 GLACIER HYDROLOGY

Glaciers and water are intimately associated, and many features are produced as a result of this interaction (Table 5). Again, thermal regime plays an important role in the routing of water through a glacier. In temperate glaciers, water flows freely over the surface, forming incised channels, and readily descends *via* **moulins** (near-vertical shafts) to the glacier bed. In valley glaciers, water generally leaves the glacier *via* a single **portal**, usually at the lowest topographic point at the glacier **snout**. Water on polythermal glaciers stays at the surface longer, and as a result **supraglacial streams** may run the whole length of a glacier and become deeply incised to depths of several metres. The sub-zero temperatures of the ice inhibit ingress of water to the interior of the glacier, but down-cutting over several years may create **englacial channels** when the ice above the stream closes up by ice creep. Otherwise, water flows towards the glacier margins, and ice-marginal streams may be extremely erosive. Outflow from polythermal glaciers tends to be along the flanks of the glacier, but much flows off the surface of the glacier or from englacial conduits at the terminus (Figure 10). Cold glaciers, **ice shelves** and **ice streams** also collect water. At sub-zero temperatures, ice can still melt from solar radiation on dust-covered or dark-coloured ice.

Lakes are commonly associated with glaciers. They occur in ice-marginal situations as **ice-dammed** lakes, on the surface as **supraglacial ponds** and at the terminus as **proglacial lakes**. Ice-dammed lakes are particularly prone to catastrophic drainage, commonly every summer. Polythermal glaciers provide particularly good sites for ice-dammed lakes, as abundant meltwater is generated in summer and continuous drainage is inhibited by cold ice. Small holes (**cryoconite holes**) and ponds may develop on a glacier surface as a result of uneven melting, and these may host a rich ecosystem of microorganisms.

TABLE 5
Terms Relevant to Glacier Hydrology

Ablation	Flocculation	Mudflow
Accumulation	Fountain	Penitentes
Adfreezing	Frazil ice	Percolation zone
Albedo	Frictional heat	Polythermal glacier
Aluvión	Geothermal heat	Portal
Arborescent drainage system	Glacial lake	Precipitate
Aufeis	Glacial lake outburst flood (GLOF)	Proglacial channel
Basal ablation	Glacial stream	Proglacial lake
Basal glide	Glacier bed	Regelation
Basal hydrology	Glacier–climate interaction	Rill
Basal melting	Glacier dynamics	Runoff
Basal shear stress	Glacier flow/motion	Sediment flow
Basal sliding	Glacier hydrology	Sediment flushing/evacuation
Biogeochemical processes	Glacier karst	Sediment plume
Blue ice area	Glacier milk	Seepage
Calcium carbonate precipitate	Glacier portal	Shelf ice
Canal system	Glacier runoff	Sliding
Candle ice	Glaciofluvial/glacifluvial sediment	Slush flow
Canyon	Glaciohydraulic supercooling	Slush zone
Cauldron collapse	Glacio-speleology	Snow swamp
Cavitation	Glaze	Subglacial cavity
Chemical weathering	GLOF	Subglacial channel/conduit
Cold glacier	Hot-water drilling	Subglacial hydrology
Cold ice	Hydraulic jacking/uplift	Subglacial lake
Conduit	Hydroelectric power generation	Subglacial meltwater channel
Crevasse trace	Hydrofracturing	Subglacial stream
Cryoconite	Ice blister	Subglacial water pressure
Cryoconite quirk	Ice-dammed lake	Supercooling
Crystal quirk	Ice-marginal channel	Supraglacial drainage
Débâcle	Ice-marginal lake	Supraglacial lake/pond
Debris (dirt) cone	Ice-shelf collapse	Supraglacial stream
Deformable bed	Icicle	Suspended sediment
Dendritic drainage	Icing	Thermal regime
Discharge	Infiltration	Transient snowline
Distributed drainage system	Linked cavity system	Upwelling (ice)
Doline	Meander	Upwelling (sediment)
Energy balance	Melt-out	Waterline notch
Englacial channel/conduit	Meltwater channel	Water-table
Epi-shelf lake	Microbial processes	Wet snow zone
Facies (basal ice)	Moat	
Fast glacier flow	Moulin	

6   GLACIAL EROSIONAL PROCESSES

Glacial erosion produces a wide range of phenomena on scales ranging from continental to microscopic, many of which are illustrated in this atlas (Table 6). Continental-scale erosional features include vast areas of **aerial scouring**, such as on the Laurentian Shield of Canada, where hills and rocky basins with lakes pervade the landscape. Smaller scale areas may be referred to as **knock-and-lochain** topography, as in the Northwest Highlands of Scotland.

FIGURE 10 Spectacular waterfall, fed by an englacial meltwater stream, emerging from the snout of Thompson Glacier's front. The sediment-rich nature of this stream indicates that it must have had a connection with the debris-rich basal layer of the glacier (July 2008).

The most impressive landforms, however, occur in mountainous regions where glaciers have scoured out deep valleys and **fjords** (Figure 11), bowl-shaped depressions high on mountain flanks (**cirques**) and narrow ridges (**arêtes**). These features occur in dissected plateau regions such as Norway, Greenland and the British Isles. They also occur in areas of alpine topography such as the Western Cordillera of North America, the Alps, the Andes and the Himalaya.

Intermediate-scale features tens of metres to a kilometre in length, such as **roches moutonnées**, are shaped by a combination of abrasion, the sandpaper effect of debris-rich ice moving across a rock surface, and **quarrying** or plucking of rock fragments by the passing ice. Some roches moutonnées may be associated downstream with the products of glacial deposition, such features being known as **crag-and-tails**.

Such features are associated with small-scale erosional features, of which linear scratch marks or **striations** a few millimetres across are the most common (Figure 12). Striations grade down into very closely spaced fine scratch marks and **polish**. Other small-scale features are a range of rock impact marks that include **crescentic gouges** and **chattermarks**. Modern glacier margins are a good place to observe the intermediate and small-scale features of glacial erosion, since here they occur on fresh bedrock surfaces that have not yet been weathered or covered in vegetation. In areas that have long since lost their glaciers, small-scale and intermediate-scale erosional landforms provide a good indication of glacier-flow directions.

In addition to features formed by direct contact between ice and bedrock, there are many forms that are the product of erosion by glacial meltwater, including **subglacial meltwater channels** and a family of metre-scale features called **p-forms**. In some areas, especially those with temperate and polythermal glaciers, meltwater may be an even more effective agent of erosion than the ice itself. This efficacy is largely the result of water mixed with sediment being under high pressure beneath a glacier, and therefore more erosive.

TABLE 6
Terms Relevant to Glacial Erosion

Abrasion	Glacial erosion	Ribbon lake
Alpine landscape	Glacial lake shorelines	Riegel
Areal scouring	Glacial landsystem	Rinnentaler
Arête	Glacial surface of erosion	Roche moutonnée
Bedform	Glacial trough	Rock basin
Bedrock step	Glaciated landscape	Rock drumlin
Bed roughness	Glacier bed	Rockfall
Boulder pavement	Glacier forefield	Rockslide
Breached watershed	Glaciotectonism/glacitectonism	Rock-slope failure
Bullet-shaped boulder	Hanging valley	Scablands
Cavetto form	Headwall	Scoop
Cavitation	Hill–hole pair	Sea loch
Chain lakes	Iceberg pits and mounds	Selective linear erosion
Channelled scablands	Iceberg plough-marks	Sichelwanne
Chattermarks	Iceberg scour marks	Sill
Chute	Ice-marginal channel	Sliding
Cirque	Joint-separation	Slope failure
Comminution	Knock-and-lochain (lochan) topography	Slot gorge
Compound cirque	Landsystem	Snow-ice avalanche pit
Conchoidal fractures	Loch	Spillway
Corrie	Meltwater channel	Stoss-and-lee form
Crag-and-tail	Moraine breach	Strandline
Crushing	Nail-head striae	Streamlined bedforms/topography
Cupola hills	Nye channel	Striae or striations
Curved (winding) channel	Nunatak	Subglacial gorge
Cwm	Overdeepened basin (or overdeepening)	Subglacial meltwater channel
Debuttressing	P-forms (or plastically moulded forms)	Tarn
Diffluence	Parabolic form	Threshold
Exfoliation	Paternoster lakes	Trimline
Finger lakes	Plucking	Trough-end
Fjord (Fiord)	Polish	Trough-head
Flyggberg	Pothole	Truncated spur
Fractures (rock)	Pressure release	Tunnel valley
Gendarme	Pyramidal peak	U-shaped valley
Glacial breach	Quarrying	V-shaped valley
Glacial buzzsaw effect	Rat-tail	Whaleback

7 GLACIAL DEPOSITIONAL PROCESSES

In combination with regions affected by glacial erosion, glacial deposition has affected a third of the world's landmass, as well as many high-latitude continental shelves. Direct glacial deposition is associated with a wide range of related processes: marine, lacustrine, fluvial and aeolian. There is therefore a huge range of features that relate to the glacial depositional environment to a greater or lesser extent; many are included in this atlas (Table 7).

As noted in the preceding text (Section 3), the manner in which deposition takes place from a glacier is strongly influenced by its disposition within the ice, and this is commonly controlled by ice structure. The textural characteristics of sediments deposited directly from the ice reflect whether it has been carried passively on the surface (**supraglacial debris**) and therefore not modified, or

FIGURE 11 The deep glacial trough of Milford Sound, South Island, is one of the finest fjords in New Zealand, and a popular tourist destination. Virgin rain forest drapes the lower walls of the fjord, and the hanging valley on the left. Mitre Peak or Rahuto (1683 m) is one of the country's most famous landmarks.

FIGURE 12 Smaller scale features of glacial erosion are striations, as on this basaltic bedrock recently exposed as a result of recession of Skálafellsjökull, Iceland (March 2015).

TABLE 7
Terms Relevant to Glacial Deposition

Ablation valley	Dump moraine	Ice-contact fan
Algae	Earth pyramids	Ice-cored moraine
Alluvial fan	Ecological succession	Ice-rafted debris
Alpine landscape	End moraine	Ice stagnation
Aluvión	Eolian sediment	Imbrication
Annual moraines	Erratic	Jökulhlaup
Bar	Esker	Kame
Basal debris/sediment	Fabric	Kame delta
Basal melting	Facet	Kame terrace
Basal till	Facies (sediment)	Kettle hole (or kettle)
Basket-of-eggs topography	Fault (sediment)	Lahar
Bedform	Flocculation	Landsystem
Bergstone mud	Flute	Lateral moraine
Blockfield	Forefield	Lateral morainic trough
Boulder clay	Furrowed lateral moraine	Latero-terminal (latero-frontal) moraine
Braid plain	Geometrical ridge network	Lava-fed delta
Braided outwash fan	Gilbert-type delta	Lodgement (lodgment)
Breach-lobe moraine	Glacial lake outburst flood (GLOF)	Loess
Bulldozing	Glacial lake shorelines	Lonestone(s)
Bullet-shaped boulder	Glacial landsystem	Megablock
Calcium carbonate precipitate	Glacial sedimentary logging	Microbial processes
Clast	Glacially transported boulder	Morainal bank
Clastic dykes	Glaciated landscape	Moraine
Cold-ice moraines	Glacier-fed delta	Moraine-mound complex
Controlled moraine	Glacier forefield	Mud
Core-logging	Glacier karst	Mudflow
Crag-and-tail	Glacier–volcano interaction	Outwash plain
Cross-bedding/cross-stratification	Glacigenic/glaciogenic sediment flow	Paraglacial processes
Cupola hill	Glaciofluvial/glacifluvial sediment	Periglacial processes
Cyclopel and Cyclopsam	Glaciolacustrine/glacilacustrine delta	Piedmont moraine
Dead ice	Glaciolacustrine/glacilacustrine sediment	Pitted outwash
Débâcle	Glaciomarine/glacimarine sediment	Precipitate
Debris apron	Glaciotectonism/glacitectonism	Proglacial area
Debris entrainment	GLOF	Proglacial lake
Debris flow	Gravel	Push moraine
De Geer moraines	Grounding-line fan	Pyroclastic deposits
Deglaciation	Grounding-zone wedge	Raft
Delta-fan complex	Ground moraine	Raised beach
Delta moraine	Hill–hole pair	Recessional moraine
Desiccation	Hummocky moraine	Response time
Diamict apron	Hyaloclastite	Ribbed moraine
Diamicton/diamictite/	Hydrofracturing	Rockfall
diamict	Iceberg pits and mounds	Rockslide
Drift	Iceberg push ridges	Rogen moraine
Dropstone	Iceberg rafting or ice-rafting	Sand
Drumlin	Iceberg turbate	Sandur (plur. Sandar)
Drumlinoid ridges or drumlinised	Ice-contact delta	Scree
ground moraine		

(Continued)

TABLE 7 *(Continued)*

Terms Relevant to Glacial Deposition

Sediment core	Subglacial debris	Till fabric
Sediment flow	Subglacial sediment	Tillite
Sediment-landform association	Subglacial till mosaic	Till plain
Sediment ripples	Subglacial traction till	Topographic inversion
Sedimentary dykes	Supraglacial debris stripe	Trough-mouth fan
Sequence stratigraphy	Suspended sediment	Turbidity flow/turbidite
Slumping	Talus	Unconformity
Spalling	Tephra	Upwelling mound
Stagnant ice	Terminal moraine	Valley train
Strandline	Thrust-block moraine	Varves
Streamlined bedforms/topography	Thrust moraine	Volcanic bomb
Stromatolites	Till	
Subaqueous sediment fan	Till delta	

whether it has come in contact with the bed where material is subject to **abrasion**, fracturing and crushing (**basal debris**). In the former case, rock material remains angular and may drape wide areas of a glacier snout until it is deposited as irregular mounds (**hummocky moraine**). In some cases, supraglacial debris may be folded, form medial moraines and be deposited in lines in the proglacial area (**supraglacial debris stripes**). Basal debris forms a poorly sorted sediment, texturally described as **diamicton**, which if deposited directly from the ice without subsequent reworking may be interpreted as **till**. Extensive sheets of till may be deposited, and may be further modified by flowing ice to form a family of landforms, referred to as **subglacial bedforms**. The most important of these are **drumlins, mega-scale glacial lineations** and **flutes**. More scattered basal or supraglacial debris is referred to as **erratics**. Analysis of the composition of tills and erratics has proved to be very useful in tracing far-distant mineral deposits.

Water in channels beneath a glacier not only creates a range of erosional features but also forms distinct depositional landforms, notably linear or sinuous ridges known as **eskers**. All these sediments are reworked to a greater or lesser extent in proximity to melting ice, by **slumping** or **debris flowage**. The rivers that emerge from a glacier also rework the sediment. With strong seasonal variations in discharge, most glacial rivers are **braided**, and sediment derived from the glacier becomes well sorted. Sands and gravels are typical of braided rivers, and finer suspended sediment may settle out in areas of slack water. In alpine regions, braided rivers may constantly swing back and forth across the floor of a valley, creating gravelly plains known as **valley trains**, such as in Alaska and the Yukon. In areas beyond the mountains that are influenced by outburst floods, wide plains with continually migrating channels called **sandar** may stretch laterally for tens of kilometres, as in Iceland.

Water bodies in contact with glaciers provide a distinctive suite of sediments (**facies**), including laminated sand and mud, disrupted by coarse debris flows and ice-rafted **dropstones**. Lake beds and tidal areas become exposed, and fine sediment is subject to wind erosion and redeposition. With all these processes ongoing, the **proglacial areas** of glaciers and ice sheets are exceedingly complex depositional environments. Extreme events are associated with glacial lakes, since they are commonly unstable. **Glacial lake outburst floods** are common in modern glacier-dominated catchments, and in the past they attained huge proportions. This is especially true in North America where they produced distinctive deeply incised terrain referred to as **channelled scablands**.

The range and relative proportions of sedimentary types are a reflection of glacier thermal regime (and thus climate) and topographic setting. Temperate valley glaciers carry a high supraglacial debris load, but most basal glacial sediment is reworked by meltwater. In contrast, many Arctic polythermal glaciers carry relatively little supraglacial debris, and extensive sheets of till

FIGURE 13 Moraines are among the most prominent of glacial depositional landforms. This example of a freshly formed moraine encloses a lake at the foot of Mount Robson (3954 m), British Columbia, Canada. This moraine is ice-cored.

may escape reworking by meltwater. Cold glaciers do not generate much of their own sediment, but rework their frozen substrate by **glaciotectonic processes**. In areas of extreme topography such as the Himalaya or Andes, the main depositional products may be derived from rockfalls.

Irrespective of the suite of sediments produced at and beyond a glacier margin, when a glacier advances and then recedes, it produces a wide variety of ridges or **moraines** (Figure 13). The processes involved include bulldozing, dumping, thrusting within and beyond the ice and other forms of glaciotectonic deformation and ice-**stagnation**. Numerous moraine types are distinguished based on position, process of formation and morphology.

Glacial deposits provide the means of determining the environmental history of an area. Dating of moraines and erratics may be challenging, but any dates obtained will facilitate the interpretation of periods of glacier advance and recession, which can then be linked to the detailed climatic records derived from sediment cores from the deep ocean, and ice cores (Section 3).

8 GLACIAL LANDSCAPES

Glacial landscapes embrace areas that are still covered by glacier ice (i.e. they are **glacierised**), and those that have previously been affected by glaciers (i.e. they are **glaciated**). Landscapes that are heavily glacierised, where the ice cover is nearly complete, include the Greenland and Antarctic ice sheets, and smaller ice masses, such as those found in the Icefield Ranges of the Alaska/Yukon border region, Svalbard, or the Patagonian Icefield and its surroundings. Other glacierised regions with a lower percentage of ice cover are the many mountain regions around the world and many Arctic archipelagos. Glacial landscapes in such areas naturally display a wide range of glaciological phenomena, as well as the erosional and depositional products of glaciers (Table 8).

TABLE 8

Terms Relevant to Glacial Landscapes (i.e. Assemblages of Erosional and Depositional Forms at the Landscape Scale)

Alpine landscape	Glacial landsystem	Periglacial processes
Antarctic oasis	Glaciated	Selective linear erosion
Basket-of-eggs topography	Glaciated landscape	Streamlined bedforms/topography
Channelled scablands	Glacier–volcano interaction	Transfluence
Deglaciation	Landsystem	
Ecological succession	Paraglacial processes	

FIGURE 14 A typical upland glaciated landscape preserving both erosional and depositional features: Borrowdale, Lake District, England.

Glaciated landscapes have been revealed as ice sheets and glaciers have receded. Since one-third of Earth's landmass was ice-covered (glacierised) during the Last Glacial Maximum (Section 2), there is now a wonderful legacy of glacial phenomena in many parts of the world, ranging from the polar regions to the tropics. Mountain glacial landscapes are the most scenic, and many of them favour tourism, although establishing an infrastructure in such regions can be challenging. The dominant features are erosional (Section 6), but depositional forms (Section 7), especially in valleys bottoms, add to the complexity of glaciated mountain landscapes (Figure 14).

Glaciated lowland landscapes tend to be dominated with a wide range of glacial depositional products, notably till, sand and gravel. Many glacial landforms are constructed of glacial materials, but in some areas glacial erosional landforms also occur. Glacial deposits commonly give rise to mineral-rich soils, and hence are beneficial to agriculture.

Glacial geologists examine these landscapes in a systematic manner, documenting especially the relationship between sediments and landforms (the **sediment/landform association**).

They also develop the concept of an integrated landscape or **landsystem**, which can be generalised to be applicable to other areas. A landsystem may link the sediments and landforms to a particular thermal type of glacier, or a particular topographic setting, as documented in this atlas.

9 GLACIERS AND HUMAN CIVILISATION

The impact of glaciers on humanity is considerable in both positive and negative ways (for relevant terms, see Table 9). The benefits of glaciers themselves may be considered to be the provision of water for irrigation, water supplies and **hydroelectric power**, and the generation of mineral-rich soils for agriculture. Glacial and **glaciofluvial sediments** of the Pleistocene age have many uses, including as an aquifer, as a source for sand and gravel in the construction industry, and as a means of locating mineral deposits. Glacial landscapes provide us with wonderful scenery, and many countries exploit these attributes for tourism. On the negative side, various hazards are associated with glaciers, including ice avalanches and outburst floods.

Of the positive attributes of glaciers that are used, irrigation is probably the most widespread. Areas of low rainfall or seasonal rainfall benefit from glacial meltwater in areas well beyond the glaciers, such as northwestern China, northwestern India and Pakistan, and coastal Peru including the city of Lima. In mountainous terrain, within the vicinity of glaciers, irrigation is provided in regions as culturally diverse as the Cordillera Blanca in Peru (Figure 15) and the Alps. Water from glaciers is also harnessed widely for hydroelectric power schemes. In the European Alps, for example, networks of dams, water intakes and tunnels supply massive hydro schemes (Figure 16). Whilst glaciers still exist, there remains considerable potential for further hydroelectric power generation, especially in developing countries.

Glacial erosion and deposition have had a profound effect on the development of agriculture. In glaciated mountain areas, agriculture is often not profitable without subsidies. Often such areas have high rainfall, so sediments, especially till and the acid soils developed on it, are often waterlogged, limiting agriculture to grazing by sheep, cattle or deer. In lowland areas, where rainfall is much lower, till provides an especially good basis for pasture, as it holds moisture well, and glaciofluvial sediments provide opportunities for crops that require well-drained soils. Such contrasts between mountain and lowland areas are especially noticeable in Britain, for example, by comparing the Lake District with the Eden Valley to the east.

Among the depositional products from glaciers, glaciofluvial sediments are the most important. Where buried, glaciofluvial sediments especially if capped with till are an excellent aquifer, providing clean drinking water to many cities. Sand and gravel aggregates of glaciofluvial

---

TABLE 9

Terms Relevant to Glaciological Impacts on Civilisation

Alpine landscape	Glacier recession (retreat)	Paraglacial processes
Aluvión	Glacier–volcano interaction	Periglacial processes
Avalanche	Glacio-eustacy	Pyroclastic deposits
Blizzard	Hydroelectric power generation	Response time
Crater glacier	Ice	Rock avalanche
Débâcle	Ice-dammed lake	Rockslide
Discharge	Jökulhlaup	Rock-slope failure
Diurnal fluctuation/variation	Lahar	Snow avalanche
Glacial lake outburst flood (GLOF)	Mass balance	Subglacial volcanism
Glaciated landscape	Moraine-dammed lake	Whiteout
Glacier–climate interaction	Mudflow	

FIGURE 15 Glacier water is widely used for irrigation in many mountain regions, as here on the slopes of Huascaran, Cordillera Blanca, Peru. It is predicted that this and similar regions will experience drastic loss of ice cover in the coming decades as a result of climatic warming.

FIGURE 16 Dam retaining, Griessee, Switzerland, constructed as part of a hydroelectric power scheme. The glacier in the background, Griesgletscher, provides most of the water, but is rapidly receding (1984).

origin are also extracted for the construction industry, notably north of the Alps, in the United Kingdom, northern Germany and Poland, and around the Great Lakes of North America. Once exhausted or no longer economically viable, sand and gravel quarries are often used for disposal of domestic and industrial waste, especially in the United Kingdom, but a full understanding of the geometry of the sediment bodies and groundwater hydrology is needed if such projects are to avoid polluting the ground.

Glaciers and their mountain landscapes provide some of the most spectacular scenery on the planet, and as such have encouraged the development of many tourist areas, commonly centred around sight-seeing, walking, climbing, skiing and other 'adventure sports'. Examples of popular and accessible glaciated mountain landscapes are Yosemite National Park in California, Glacier National Park in Montana, the Banff and Jasper National Parks in the Canadian Rockies, the Alps of central Europe, the national parks in the Southern Alps of New Zealand, the fjords of western Norway, the Snowdonia National Park in Wales, the Lake District National Park in England, and the Scottish Highlands. Less accessible areas are reached by increasingly popular marine cruises, such as into the fjords of southern Alaska, and around the coasts of Svalbard, Greenland, and the Antarctic Peninsula.

On the negative side, glaciers have been responsible disasters of several different types. For example, in the worst affected region, the Cordillera Blanca in Peru, over 32,000 people lost their lives in the twentieth century to glacier-related disasters. Whilst not on the same scale as tsunamis, volcanic eruptions or earthquakes, glaciers are associated with ice **avalanches** and outburst floods, which not only cause loss of life, property, agriculture and infrastructure but also may result in a major setback for countries with fragile economies.

The worst recorded single glacier-related disaster was an ice avalanche that was triggered by an earthquake of 7.8 magnitude on the Richter scale. This took place on Nevado Huascarán (6768 m) in the Cordillera Blanca, Peru, on 31 May 1970. Part of the summit glacier and the bedrock beneath it fell away, swept down the mountain, mixing with moraine and water from a lake *en route*. The 50 million cubic metres of rock, ice and debris travelled 16 km in three minutes to the valley bottom, burying the town of Yungay and killing 18,000 people (Figure 17). A further 52,000 people lost their lives in this earthquake. Other devastating ice avalanches have included one that took place in 1962 from the same source (4000 fatalities), one that resulted from the collapse of the Allalingletscher in Switzerland in 1965 (88 fatalities), and another that swept through the village of Karmadon in the Caucasus Mountains in 2002 (120 fatalities). Ice avalanches are difficult to predict if triggered by external forces such as earthquakes, but since the Allalingletscher disaster, Swiss glaciologists have developed a capability of identifying potential avalanche sites, and monitoring and predicting their onset.

Almost equally devastating are **glacial lake outburst floods**, commonly referred to as GLOFs. These are the result of the failure of moraines, often ice-cored, which retain lakes that grow as debris-mantled glaciers recede and thin. Displacement waves from rockfalls and ice-cliff **calving**, melting of the ice core, heavy rain or seepage can all affect the integrity of the moraine dam. Many such dams are 100 m high or more, and the number of such lakes is increasing due to glacier recession. The Cordillera Blanca in Peru was once again the site of a major disaster when, in 1941, an outburst flood hit the town of Huaraz, and over 6000 people lost their lives. Other moraine lake dams in the region have also failed, but several of the most hazardous ones have been remediated by lowering their spillways or excavating diversionary tunnels. The Himalayan–Karakorum Range has many potentially hazardous lakes, but even following several failures, and loss of life and infrastructure, remediation measures have been highly selective.

Another type of outburst flood results from the sudden draining of an **ice-dammed lake**. This type of flood is often seasonal. In the Alps and Western Cordillera of North America, the lakes typically fill up during the summer until the head of water is sufficient to lift the damming glacier

FIGURE 17 Of the many glacier hazards, ice avalanches mixed with debris and water have proved to be the deadliest. This photograph from 1980 shows the full route of the Huascaran avalanche (shaded), which was triggered by a powerful earthquake on 31 May, 1970. Ice falling from the summit icefield mixed with morainic debris and water, creating a fast-flowing debris-flow that rushed over an intervening ridge and destroyed the town of Yungay, situated on the large fan.

and force its way underneath the glacier. The resulting tunnels close up in the following winter by ice deformation, allowing the lakes to reform. Close monitoring of such lakes tends to prevent disasters. The well-known Icelandic floods known as *jökulhlaups* are the result of ice-constrained lakes growing over a geothermal field or an active volcano, and then being released catastrophically. These too are closely monitored, but commonly such floods breach the highway that crosses southern Iceland.

Glacier hazards can be classified as potential low-frequency, high-impact events. They often occur in remote mountain regions at high altitude, which are challenging to reach, and even more difficult to remediate. Nevertheless, some hazards, such as floods, may have consequences hundreds of kilometres downstream. Satellite imagery provides the means of identifying hazards in remote areas, but verification on the ground is still necessary. The skills of glaciologists and glacial geologists are required to evaluate the complex interaction of processes in these regions, many of which are illustrated in this atlas. However, developing countries, sadly, often lack the resources to fully monitor and remediate these hazards.

10   IMAGE SELECTION

We have selected the images for this atlas with a view to:

1. Providing coverage that is reasonably global
2. Illustrating some features from both the ground and from the air
3. Using freely available satellite imagery from NASA to show the larger scale features
4. Providing matching pairs of photographs for readers in both North America and Europe
5. Including examples from Southern Hemisphere continents and Asia that will be familiar to readers in those regions
6. Giving a strong emphasis to Antarctica, which contains by far the largest volume of ice on the planet

We appreciate that there are some significant gaps in our geographical coverage, notably the Russian Federation and surrounding states, but our preference has been to use our own photographs where possible, in order to exercise control on quality.

The selection of images from our extensive photographic collections has primarily focused on how clearly a particular feature is represented. Secondarily, we have then tried to achieve a reasonable balance between countries where interest in glacial environments is strong. The number of images in this volume for each major region of the world is summarised in Figure 18.

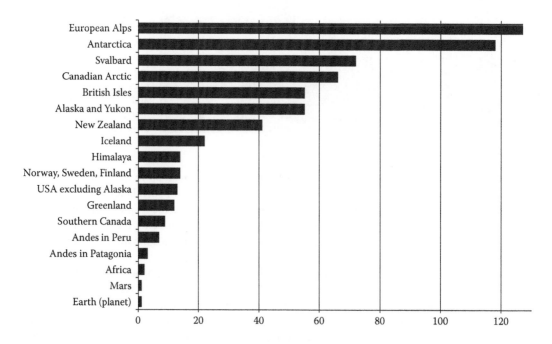

FIGURE 18   Geographical distribution of photographs in this volume.

11 SOURCE MATERIALS

In this book, we attempt to describe briefly and illustrate most of the phenomena associated with glaciers themselves and their historical legacy. We regard this volume as complementary to several textbooks and popular science volumes, including:

Alean, J. and Hambrey, M.J. 2013. *Gletscher der Welt*. Haupt AG. A highly illustrated large-format, popular-science-level account, presented as a tour of the world's glaciers, with personal reminiscences of two glaciologists. (Also published in French as *Tour du Monde des Glaciers*.)

Benn, D.J. and Evans, D.J.A. 2010. *Glaciers and Glaciation, 2nd Edition*. Hodder Education. An impressive comprehensive textbook covering glaciology, glacial geology and glacial geomorphology, pitched at advanced undergraduate and postgraduate levels.

Bennett, M.R. and Glasser, N.F. 2010. *Glacial Geology: Ice Sheets and Landforms, 2nd Edition*. Wiley. Another textbook covering glaciology, glacial geology and glacial geomorphology, but with plenty of case studies, it is especially suited to middle and senior undergraduates.

Cuffey, K.M. and Paterson, W.S.B. 2010. *The Physics of Glaciers*. Butterworth-Heinemann (Elsevier). A rigorous university-level textbook covering the mathematical and physical aspects of glaciers.

Hambrey, M.J. and Alean, J. 2004. *Glaciers, 2nd Edition*. A popular science account of glaciers and their landforms, richly illustrated with the authors' photographs, suitable for secondary/high school students and junior undergraduates, as well as laypeople, Cambridge University Press.

Knight, P.G. (Ed.) 2006. *Glacier Science*. Blackwell Publishing. Many aspects of glaciology and glacial geology/geomorphology, presented in short articles by many of the world's experts.

Lowe, J. and Walker, M. 2015. *Reconstructing Quaternary Environments*. Routledge (Taylor & Francis). A comprehensive evaluation for undergraduates and researchers of the most recent period of geological time, embracing especially the glacial record.

Singh, V.P., Singh, P. and Haritashya, U.K. (Eds.) 2011, *Encyclopedia of Snow, Ice and Glaciers*. Springer. A large compendium of most aspects of glaciers and their products, written by many of the world's leading glaciologists.

We have consulted the preceding volumes to identify and verify many of the terms in this volume. In addition, we have consulted older literature, numerous original papers and the Internet for some additional terms, including:

Armstrong, T., Roberts, B. and Swithinbank, C. 1973. *Illustrated Glossary of Snow and Ice*. Scott Polar Research Institute.

Embleton, C. and King, C.A.M. 1975. *Glacial Geomorphology*. Arnold.

Hambrey, M.J. 1994. *Glacial Environments*. UCL Press and UBC Press.

Menzies, J. (Ed.) 1995. *Modern Glacial Environments. Vols. 1 & 2*. Butterworth-Heinemann.

Sugden, D.E. and John, B.S. 1976. *Glaciers and Landscape*. Arnold.

Working Group on Mass-balance Terminology and Methods of the International Association of Cryospheric Sciences (IACS) 2011. *Glossary of Mass Balance and Related Terms*. IHP-VII Technical Documents in Hydrology No. 86. http://unesdoc.unesco.org/images/0019/001925/192525E.pdf

Most of the photographs in this volume (summarised by region in Figure 18) have been taken by the authors over a period of 45 years. However, for the features that we have not been able to observe and photograph, we are grateful to our colleagues for filling in the gaps; they are credited in the appropriate places. The diagrams are mostly our own, but sometimes simplified from figures in research papers.

12   HOW TO USE THIS ATLAS

The emphasis in this atlas is on features that can be observed in the field, and to a lesser extent from the air and from space. We have attempted to be reasonably comprehensive for terms related to glaciology, glacial geology and geomorphology. We have also included a selection of key terms describing sea ice and periglacial phenomena, but only where they are closely associated with modern glaciers. Readers will need to consult other works (listed in Section 11) to achieve comprehensive coverage of these related environments. Some terms listed are of a conceptual nature and so do not have accompanying photographs, but in some cases we include diagrams to help illustrate the point.

Many terms cannot be defined without reference to other terms, so to facilitate cross-referencing entries are printed in bold letters. In some cases, broad terms embrace a multitude of different forms, some examples being **crevasse**, **glacier** and **moraine**. In such cases, a general description is provided without an illustration and the reader will be able to cross-reference to the subsidiary terms arranged alphabetically, for example, **splaying crevasse**, **valley glacier** or **hummocky moraine**.

Lastly, given the increasing awareness of how quickly glaciers on the whole are receding in response to climatic change, we have given the month and year in which those photographs featuring glaciers were taken. In many cases, a return to those glaciers will reveal that considerable changes have taken place since.

A

A

Ablation

The process of wastage of snow or ice by melting, wind erosion and loss into a body of water (**calving**). The terms **ablation area** and **ablation zone** refer to that part of a glacier surface, usually at lower elevations, over which ablation exceeds accumulation (Figure A.1a). The **ablation season** is that period of the year when there is net loss of ice from a glacier. In most regions, the ablation season coincides with the summer months (Figure A.1b), but special cases are where ablation and accumulation occur at the same time, or when ablation occurs all year round. Where accumulation occurs down to sea level, as in many parts of Antarctica, ablation is dominated by calving.

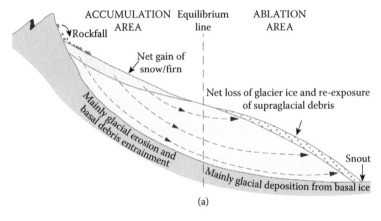

(a)

(b)

FIGURE A.1 (a) Schematic longitudinal section through a cirque or valley glacier illustrating accumulation and ablation areas in the context of glacier flow. (b) An ice cap on Vega Island, NE Antarctic Peninsula, undergoing heavy melting above volcanic cliffs (February 2010). Most glaciers in this region are undergoing rapid recession in response to regional warming.

A

Ablation area crevasses

An assemblage of V-shaped open fractures that form in bare ice in the **ablation area** of a glacier when freshly formed crevasses are typically few tens of metres deep. Melting results in rounding of the edges of the crevasses and eventually a high-relief surface topography of ice pinnacles and ridges (Figure A.2).

FIGURE A.2   Ablation area crevasses in Skalafellsjökull, Iceland (March 2015).

Ablation valley

Obsolete term for lateral morainic trough.

Abrasion

The wear of rock surfaces by a sliding glacier through polishing and scratching by embedded debris (Figure A.3). The process is commonly likened to the effect of sandpaper and produces **striations**.

FIGURE A.3   Glacially abraded granite surface at the snout of Rhonegletscher, Switzerland. The polished and striated surface, reflecting the sunlight, is almost shiny.

Accreted/accretion ice

Ice that is formed by the freezing of water at the base of an ice mass (Figure A.4).

FIGURE A.4 Freeze-on of supercooled subglacial water, ascending the adverse slope of a subglacial overdeepened basin, leads to the formation of debris-rich basal ice at Svínafellsjökull, Iceland. (Photograph by Simon Cook.)

Accumulation

The processes that contribute to the mass of a glacier, typically by the building up of a pack of snow, firn, refrozen slush and meltwater. Net accumulation for one year is the material left over at the end of the melt-season. The **accumulation area** or **zone** is that part of a glacier surface, usually at higher elevations, on which there is net accumulation of snow and other contributors to mass (Figures A.1b and A.5). As a result of burial, snow is transformed into **firn** and then **glacier ice**. The **accumulation area ratio** (AAR) is the ratio between the accumulation area and the total area of the glacier, expressed as a percentage or on a scale of 0 to 1. When glaciers are in a steady-state condition, the AARs typically range from 0.55 to 0.65.

FIGURE A.5 Annual layers of snow and firn exposed in the wall of a crevasse on Weissmies, Switzerland. These layers are typical of the accumulation area of a glacier. The thin dark layers represent summer surfaces where some melting has occurred and dust concentrated (Summer 1982).

Adfreezing

The process whereby water is frozen onto the base of a glacier as **accretion ice** (Figure A.4), usually as a result of circulation of cold air from outside, or by penetration of a winter cold wave through the ice.

Advance (see **glacier advance**)

Aeolian sediment

Wind-blown sediment that is found in cold arid regions where the temperature remains below zero for most of the year. Sediment size fractions are mainly of sand but may also involve granules and pebbles as a result of movement by strong winds (Figure A.6a). Aeolian sediment is found on the surface of glaciers and as drapes and dunes beyond the ice margin in cold arid regions such as the Dry Valleys of Antarctica (Figure A.6b). In temperate regions, wind-blown sediment is commonly silt-grade, and is referred to as **loess**.

(a)

FIGURE A.6 (a) Wind-blown sand with ripples on the inner flanks of a moraine at Imja Glacier, Khumbu Himal, Nepal. (b) Wind-rippled sand banked against the ice cliffs of Wright Lower Glacier, Dry Valleys, Antarctica (January 2010).

(b)

Air bubbles

Glacier ice contains a large volume of air, trapped as snow turns to ice by compaction. The air becomes concentrated in discrete, closely spaced air bubbles, which impart a pale blue tone to the ice. Air bubbles are commonly ellipsoidal in shape as a result of internal deformation. Air bubbles are also trapped in water (**regelation**) ice that freezes to the base of a glacier (Figure A.7). In an **ice core**, air bubbles contain representative samples of the atmosphere when they were trapped, and have been used to determine climatic changes through time.

FIGURE A.7 Air bubbles trapped in the basal ice zone of Stagnation Glacier, Bylot Island, Canadian Arctic (July 2014).

Albedo

A measure of reflectivity of solar radiation from snow or ice surface. Specifically, albedo is the ratio of the reflected radiation flux density to the incident flux density. Albedo values of fresh snow are in excess of 80%, but those are significantly less for exposed glacier ice, and even less for debris-covered ice or surface ponds (Figure A.8).

FIGURE A.8 Contrast in albedo between clean glacier ice forming pinnacles and sand-covered ice forming depressions on the surface of Wright Lower Glacier, Victoria Land, Antarctica. The sand has absorbed more radiation and lowered the surface more quickly than for the highly reflective clean ice (January 2010).

Alga (plural algae)

A simple, single-cell plant belonging to the kingdom Protista that is usually found in wet environments. The blue-green variety is common in stagnant pools close to, and beyond, the margins of a glacier (Figure A.9). Red algae occur on snow in the accumulation area of a glacier, especially in the polar regions. Algae are most obvious from mid to late summer. See also **microbial processes**.

FIGURE A.9 Blue-green algae formed in a small pond in recently exposed ground close to Fountain Glacier, Bylot Island, Canadian Arctic.

Alluvial fan (see **braided outwash fan**)

Alpenglow (from the German *Alpenglühen*)

The rosy glow seen on high mountains at sunset or sunrise. This optical phenomenon is most pronounced when the mountains are covered by snow and ice, as in alpine regions (Figure A.10).

FIGURE A.10 Alpenglow at sunset on the summit and hanging glaciers of Aoraki/Mount Cook, Southern Alps, New Zealand (April 2011).

Alpine glacier (*cf.* **valley glacier**)

A self-contained glacier that flows through mountainous terrain, typically occupying a valley (Figure A.11).

FIGURE A.11   Typical alpine (or valley) glacier: Grosser Aletschgletscher, the longest in the Swiss Alps. It flows from several 4000 metre peaks from right to left (2010).

Alpine landscape

A predominantly erosional landscape of mountains, **cirques**, **glacial troughs** and **fjords** carved out by glacier ice (Figure A.12). The mountains are typically jagged, and the valleys and fjords are steep-sided with deep basins.

FIGURE A.12   Alpine landscape of jagged peaks, ridges, steep-sided valleys and fjords, near Reine, Lofoten Islands, Norway.

A

Aluvión (from Spanish *Peruvian*)

A catastrophic flood or debris flow, commonly resulting from the failure of a moraine-dammed lake (Figure A.13).

FIGURE A.13   Huge granite boulders deposited during the aluvión which damaged parts of the city of Huaraz, Cordillera Blanca, Peru, in 1941. (Photo by Arnold Heim; from 'Wunderland Peru', Hans Huber Verlag Bern.)

Annual moraines

A collection of small ridges of debris, pushed up annually by winter advances of a glacier during a period of recession. Spacing is usually a few metres, recording the net retreat each year (Figure A.14).

FIGURE A.14   Annual push moraines formed a few decades ago by Worthington Glacier, Alaska. Vegetation is already colonising the terrain (June 2012).

Antarctic oasis (see oasis)

Arborescent drainage system (see dendritic drainage system)

Arcuate crevasses

Deep-open fractures curving across a glacier as arcs. The fractures are usually concave facing down-glacier and form as the glacier approaches a steep slope. As flow progresses, they become transverse and then convex facing down-glacier. They may attain lengths of a few hundred metres and depths up to 30 metres in alpine valley glaciers (Figure A.15).

FIGURE A.15 Looking down on arcuate crevasses in Vadret Pers, Switzerland. Ice is flowing from left to right, and the crevasses form as the bed steepens. An ice-fall follows as the ice surface becomes increasingly fractured (September 2012).

Arcuate (transverse) foliation

Layers of intercalated coarse bubbly (white) and coarse clear (dark) ice, produced by deformation as the ice reaches the base of an icefall. The original layers are fractures (closed crevasses, **crevasse traces**) that have been strongly compressed. The arcuate form is the result of faster flow in the middle of the glacier (Figure A.16a). Internal deformation also rotates the initial near-vertical structure into a gently inclined state at the apex of the arcs (Figure A.16b).

(a)

FIGURE A.16 (a) Arcuate foliation and associated ogives on Root Glacier, Wrangell Mountains, Alaska, demonstrating faster flow in the middle of the glacier as it flows from left to right (July 2011). *(Continued)*

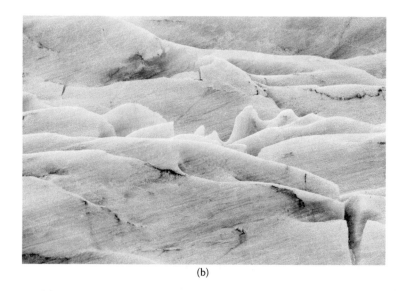

(b)

FIGURE A.16 (Continued) (b) Telephoto of arcuate foliation in Svínafellsjökull, Iceland. The foliation in this outlet glacier from the ice cap of Vatnajökull forms in an icefall as a vertical structure, which then rotates during flow to a gently dipping attitude by the time it reaches the snout, as here. The main ice types, coarse bubbly (light blue) and coarse clear (dark blue), are visible. The crevasses allow the three-dimensional form to be observed (March 2015).

Areal scouring

Large-scale erosion of bedrock in lowland areas by an ice sheet. The resulting landscape consists of smoothed ice-worn bedrock forming small hills and water-filled bedrock hollows (Figure A.17).

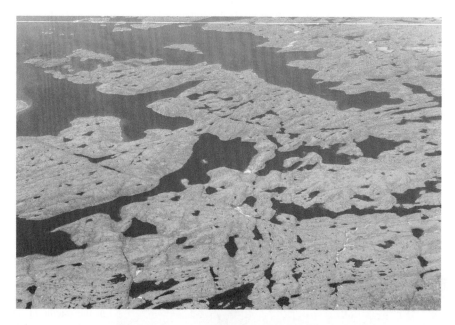

FIGURE A.17 Areal scouring from the air in southern Baffin Island. Ice flow in this photograph was from right to left, and preferentially exploited bedrock structure, hence the pronounced linear features.

Arête (from the French *Arête*)

A sharp, narrow ridge, formed as a result of glacial erosion from both sides, commonly where the headwalls or side walls of cirques are eroded backward until they meet (Figure A.18).

FIGURE A.18   Arête forming the Snowdon Horseshoe in North Wales. The prominent ridge, Y Lliwedd, is seen from the summit of Snowdon, with the cirque containing Llyn Llydaw to the left. The prominent path forms part of one of the finest scrambling routes in Britain, the Snowdon Horseshoe.

Ash-layering

In regions of volcanic activity, the accumulation of snow may be supplemented by ash from eruptions, eventually forming a stratified mass of glacier ice and ash layers that deforms during glacier flow (Figure A.19). Ash-layering has been used to provide a dateable record of volcanic eruptions.

FIGURE A.19   Ash-layering in a small glacier within the main crater of the active volcano Ruapehu, North Island, New Zealand (April 2008).

A

Aufeis (from the German *Aufeis;* also known as **icing**)

Ice that accumulates in sheets during the winter along streams and rivers, usually in the Arctic or sub-Arctic (Figure A.20a). Aufeis originates from ground water and glaciers that continues to flow even after the summer melt-season ceases. Discharge may be discontinuous, as a result of which several successive layers of ice build up, sometimes to thicknesses of several metres. Each layer comprises vertically orientated crystals known as **candle ice**, which fall apart in the following summer when streams cut through the Aufeis (Figure A.20b). Not all accumulations of Aufeis melt each summer, and those may be buried by glacial outwash gravel, becoming preserved as a form of ground ice.

(a)

(b)

FIGURE A.20   (a) Aerial view of Aufeis at Fountain Glacier, Bylot Island, Canadian Arctic. (b) Multiple layers of Aufeis, 3–5 metres thick, in front of the glacier, showing candle-like crystal form. The mid-summer melt generates an active glacial river that undercuts the Aufeis, causing it to collapse (July 2014).

Avalanche (ice; snow)

A fall, flow or slide of snow or ice that suddenly becomes detached from a steep mountainside (snow) or hanging glacier (ice) (Figure A.21). As the avalanche progresses downslope, it sometimes picks up vegetation, rocks and soil. The length an avalanche travels is known as the **run-out distance**. Ice and snow avalanches, combined with rock, are a major component of the accumulation of high-mountain glaciers, such as those in the Himalaya, and result in **debris-mantled glaciers**.

FIGURE A.21 An ice avalanche caused by the collapse of a steep ice face at the edge of a recently exposed cliff at Unterer Grindelwaldgletscher, Switzerland (September 2009).

Avalanche cone (avalanche apron)

A steep cone-shaped accumulation of snow, ice and rock debris that accumulates on the steep rocky flanks of a glacier. Where several cones coalesce, they form an **avalanche apron** (Figure A.22).

FIGURE A.22 A series of avalanche cones. Merging together to form an avalanche apron, along the flanks of Gaißbergferner (also known as Gaisbergferner), Tirol, Austria (September 2014).

Avalanche deposit

The aftermath of an ice avalanche is a pile of ice (and sometimes rock) debris. The distance travelled is the run-out distance, which may extend far beyond the source ice cliff (Figure A.23).

FIGURE A.23 An ice avalanche from a hanging glacier on the south face of the Mönch, Switzerland. The ice avalanche descended over a smooth area in a firn basin of Grosser Aletschgletscher, producing a grooved channel bounded by levées (July 1984).

Avalanche-fed glacier

A glacier or glacier tongue, which is sustained by ice avalanches from above. The source is usually a collapsing ice cliff perched above a steep rock slope, and the two ice masses are physically separated. The ice accumulates as a cone or series of cones below the cliff (Figure A.24).

FIGURE A.24 Ice avalanches falling from a hanging glacier contribute to the accumulation of part of Victoria Glacier, Lake Louise, Banff National Park, Canada (June 2006).

B

Balanced budget

The state of a glacier in which gains and losses of mass have been approximately equal for a number of years; that is, the mass balance has remained close to zero. Glaciers with a balanced budget are steady state.

Band ogives (see ogives)

Bar

Part of the bed of a braided river, typical of glacial environments, which is elevated compared with adjacent channels (Figure B.1). Bars are predominantly composed of sand and gravel deposited during high-discharge events. The particles in a bar are typically graded from gravel to sand in a downstream direction. Numerous bars characterise glacial rivers, and those migrate constantly during flood events. See also **glaciofluvial sediment**.

FIGURE B.1   A typical bar in the braided river fed by two Swiss glaciers, Vadret da Roseg and Vadret da Tschierva, in 2014. Note the coarsening downstream nature of the sediment on the bar and the elongated, pointed morphology.

B

Basal ablation

The removal of ice at the base of a glacier by melting. Frictional heat from sliding, geothermal heat, erosion by subglacial streams and air circulation play a role in melting the base of a glacier, which results in mass loss (Figure B.2). Melting also occurs if the ice is subject to high pressure (pressure melting), which is generally balanced by refreezing (**regelation**).

FIGURE B.2   Wet, melting and dripping ice in a cave at the base of the Rhonegletscher, Switzerland, in summer 2004.

Basal accretion

The freezing of marine ice onto the base of an ice shelf. This type of ice (referred to as **frazil ice**) is formed of vertical elongate crystals that are clear or opaque with a greasy texture (arising from the salt content). Sometimes frazil ice which forms on the sea bottom traps organisms, so these too are incorporated into the ice shelf. If there is net surface ablation of the ice shelf, frazil ice may become the main constituent in parts of the ice shelf, and the organisms can melt out at the surface (Figure B.3).

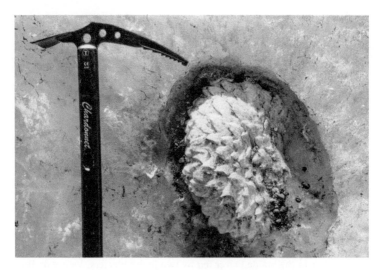

FIGURE B.3   The surface of part of the southern McMurdo Ice Shelf, Antarctica, is subject to so much net ablation that basally accreted marine ice reaches the surface, exposing marine organisms such as coral, previously trapped in frazil ice (December 2010).

Basal accumulation

The gain in the mass of ice at the base of a glacier by freezing of water. The accreted ice is typically made of coarse clear ice crystals with varying proportions of debris, also known as **basal ice**. Basal accumulation also includes snow that has been blown into caves and channels at the base of a glacier, where it is transformed to ice by percolating meltwater and deformation (Figure B.4).

FIGURE B.4 White coarse clear ice at the base of Fountain Glacier, Bylot Island, Canadian Arctic. This originated from snow, blown into cavities beneath the glacier, and then altered to ice by deformation (simple shear) at the interface between glacier ice and the bed (July 2014).

Basal crevasse

A crevasse that propagates upwards from the bed of a glacier by several metres (Figure B.5). Groups of basal crevasses are commonly linked to glacier surging when sediment may be squeezed up into them, leaving behind **crevasse-squeeze ridges**. Basal crevasses are also believed to form beneath floating **glacier tongues** and **ice shelves**, where they facilitate calving. Basal crevasses have rarely been observed directly, however.

FIGURE B.5 A basal crevasse in the true left margin of actively surging Comfortlessbreen, northwestern Spitsbergen, Svalbard. Subglacial debris has been squeezed up into the open space (July 2009).

Basal debris/sediment

Debris that is incorporated at the base of a glacier. It includes rock fragments, ground-up bedrock (**rock flour**), as well as previously deposited sediment (Figure B.6). In a wet-based glacier, basal debris is modified by contact between ice and the bed, is poorly sorted and shows partial rounding and scratching of stones. In a cold-based glacier, the glacier substrate may be incorporated *en masse* into the base without much modification (*cf.* **subglacial debris**).

FIGURE B.6   Richard Waller examines basal debris at the base of a small unnamed glacier near Fountain Glacier, Bylot Island, Canadian Arctic (July 2014).

Basal glide

Deformation of an ice crystal internally along discrete planar surfaces called **basal planes** (Figure B.7).

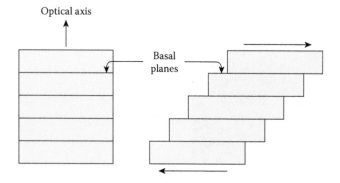

FIGURE B.7   Illustration of slip along basal planes within an ice crystal and their relationship with the optic-axis or *c*-axis. (Modified from the Centre for Ice and Climate, University of Copenhagen.)

Basal hydrology (see **subglacial hydrology**)

Basal ice layer

The assemblage of layers of ice at the bed of a glacier that is the product of melting and refreezing (**regelation**). It is strongly layered, sheared and incorporates a variable amount of debris (Figure B.8a and b).

(a)

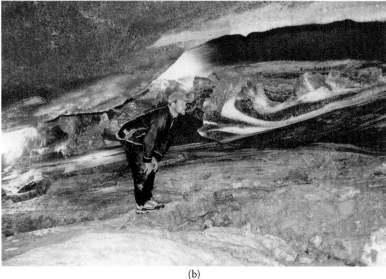

(b)

FIGURE B.8 Basal ice layer in temperate glaciers in Switzerland. (a) The dark ring around the snout of Gornergletscher marks the basal ice layer of this glacier (August 2011). (b) Glaciologist Bryn Hubbard examines the deformed basal ice layer in a cave beneath Glacier de Tsanfleuron (August 2000).

Basal melting (see **basal ablation**)

Basal plane

The plane within the hexagonal ice crystal that is normal to the optic axis. See **basal glide** (Figure B.7).

Basal shear stress

The force exerted by an ice mass on its bed.

Basal sliding

The sliding of a glacier over bedrock – a process usually facilitated by the lubricating effect of meltwater (Figure B.9). See also **glacier flow**.

FIGURE B.9   The sliding process illustrated in the base of an unnamed glacier on the south-western flank of Nevado del Pisco in the Peruvian Andes. The older icicles are tilted, whilst newly formed ones have not yet rotated (1978).

Basal till

Poorly sorted sediment that is deposited directly from the basal ice layer of a glacier (Figure B.10). Deposition is by tractional processes at the glacier base (*cf.* **lodgement till**, **melt-out till**, **subglacial traction till**).

FIGURE B.10   Basal till containing large boulders exposed above the north shore of Loch Torridon, Northwest Highlands, Scotland. Geological hammer beneath largest boulder for scale.

Basket-of-eggs topography

An extensive area of small streamlined hills of basal glacial sediment called **drumlins** (Figure B.11). Such areas are indicative of fast glacier flow.

FIGURE B.11 Low aerial perspective of basket-of-eggs topography, showing several drumlins, near Gossau, Kanton Zürich, Switzerland. The view is towards the southeast, against the flow direction of Linth Glacier during the Last Glacial Maximum.

Bedform

A broad term to describe a variety of morphological features that formed at the base of a wet-based ice mass and are now exposed on land, or in a shallow sea or lakes. They include both erosional and depositional forms, which are described separately. Erosional features in bedrock include **roches moutonnées** and streamlined bedrock, and depositional forms are **drumlins** and **mega-scale glacial lineations**.

Bedrock step

A steep reach in a longitudinal profile resulting from uneven glacial erosion. The step may coincide with resistant bedrock. The top surface becomes smoothed and striated, and the steep face quarried or plucked. With a glacier occupying the valley, the bedrock step is usually associated with an **icefall** (Figure B.12).

FIGURE B.12 In the upper reaches of Grosser Aletschgletscher, the tributary of Ewigschneefeld passes over a bedrock step, creating an icefall with numerous crevasses and wave ogives below (Summer 2004).

Bedrock raft (see **raft**)

Bed roughness

The vertical variation of a surface, with the horizontal variation along a measured profile, such as that beneath an ice sheet or valley glacier (Figure B.13). Bed roughness is affected by preglacial topography, geological structure, thermal regime at the base, ice dynamics, glacial erosion and subglacial sediment characteristics. Roughness can be quantified using a fast Fourier transform analysis. Roughness has been used to reconstruct past ice-sheet dynamics and to characterise the bed beneath the Antarctic Ice Sheet using radio-echo sounding techniques. Bed roughness can also be applied to valley glaciers.

FIGURE B.13 The recently exposed limestone bedrock at Glacier de Tsanfleuron, Switzerland, demonstrates how rough the bed of a glacier can be on a scale of several metres. The glacier lies immediately to the right of this scene (Summer 2000).

Benchmark glacier (see **index glacier**)

Bergschrund (from the German *Bergschrund*)

An irregular open fracture (**crevasse**), usually running across an ice slope in the accumulation area of a glacier. A bergschrund forms where active glacier ice pulls away from ice adhering to the steep mountainside (Figure B.14).

FIGURE B.14 Climbers travel below and parallel to a prominent bergschrund high above Oberer Grindelwaldgletscher, Switzerland (Summer 2004).

Bergstone mud

Homogeneous sandy mud with scattered stones, deposited mainly from suspension and from icebergs in a glacier-influenced fjord or bay (Figure B.15).

FIGURE B.15 Bergstone mud exposed at low tide at Hamilton Point, James Ross Island, Antarctic Peninsula. Note also the iceberg scours (January 2002).

Bergy bit

A piece of floating glacier ice, generally regarded as not exceeding 5 m above the waterline, and not more than 10 m across (Figure B.16). Bergy bits are produced where glaciers calve into a lake or the sea.

FIGURE B.16 Stranded bergy bits at low tide on the shore of Ossian Sarsfjellt, northwest Spitsbergen (Svalbard). The source tidewater glacier is Conwaybreen, behind the photographer.

Bergy seltzer

The effect of the release of pressurised air from melting glacier ice in a drink, such as whisky (or whiskey), usually manifested by a fine spray on one's face.

Biogeochemical processes

The close interaction between chemical and biological processes. As applied to glaciers, dust and sediment at the glacier surface facilitate the growth of microbes and are commonly associated with **cryoconite holes**. Biogeochemical processes are also prevalent beneath a glacier (Figure B.17), and on recently exposed ground, where weathering of bedrock is also facilitated by microbial activity.

FIGURE B.17 Blood Falls at the terminus of Taylor Glacier, Victoria Land, Antarctica. The falls are the product of saline discharge of iron-rich waters from beneath the glacier. The brine hosts a viable ecosystem of microbes, and is a good illustration of the interaction of geochemical and biological processes (November 2006).

Blizzard

A severe weather condition characterised by high winds, falling or blowing snow and poor visibility. The US National Weather Service defines a blizzard as having wind speeds of more than 35 mph (15.6 m/sec), and visibility of less than 0.25 mile (402 m) for at least 3 h. The UK Met Office defines a blizzard as 'moderate or heavy falling snow (either continuous or in showers) with wind speeds of 30 mph (13.4 m/sec) or more and a reasonably extensive snow cover'. Blizzards are a feature of glacier weather, leading to near zero visibility and potentially total disorientation, especially when thick enough to induce **whiteout** conditions (Figure B.18).

FIGURE B.18 Blizzard at a field camp on the southern McMurdo Ice Shelf, Antarctica. It is nearly obscuring the tracked vehicle and glaciologist from a distance of a few metres.

Blockfield (see **Felsenmeer**)

Blue ice

Glacier ice lacking air bubbles that has the appearance of being blue. Ice crystals absorb all colours except the blue part of the visible spectrum. Blue ice forms as a result of recrystallisation and is usually visible in ice cliffs at the terminus of **tidewater** or **freshwater glacier**, and in some **icebergs** or **bergy bits** (Figure B.19).

FIGURE B.19 A bergy bit comprising blue ice derived from the outlet glacier Breiðamerkurjökull, floating in the lagoon of Jökulsárlón, Iceland (March 2015).

Blue ice area

An area of exposed glacier ice, the term usually referring to the Antarctic Ice Sheet. They represent areas of net ablation, induced by lack of snowfall combined with sublimation enhanced by strong winds. The surface is typically polished and scalloped (Figure B.20).

FIGURE B.20 A blue ice area high up on Shackleton Glacier, which flows from the East Antarctic Ice Sheet into the Ross Ice Shelf. The scalloped, polished surface is hard and the scallops have sharp rims. This ice is difficult to walk on without crampons (February 1996).

Boudin/boudinage (from the French for sausage)

A sausage-shaped block of less ductile material separated by a short distance from its neighbours within a more ductile medium. Where different ice types or layers of sediment are involved, the term 'competence-contrast boudinage' is used, with fine-grained ice or sediment-rich ice typically being less ductile than surrounding coarse bubbly ice (Figure B.21a). These boudins normally form perpendicular to the maximum compressive strain, but if rotated a shear component may also be involved. Some boudins occur within a mass of ice with strong deformation layering (**foliation**), and are referred to as foliation boudinage (Figure B.21b). Both asymmetric and symmetric varieties occur, depending on whether paired boudins are offset or not.

(a)

(b)

FIGURE B.21   The two main types of boudinage in glacier ice. (a) Competence-contrast boudinage in the basal zone of Wright Lower Glacier, Victoria Land, Antarctica. The boudins of sand with interstitial ice are separated and rotated within coarse bubbly ice (January 1999). (b) Foliation boudinage, showing detail of pinching at the neck, in Gaißbergferner, Tirol, Austria (September 2014).

Boulder clay (synonym: till)

The deposit left by a glacier or ice sheet that is typically poorly sorted and contains glacier-abraded stones. Since not all documented 'boulder clays' contain clay or boulders, this English term is no longer favoured by glacial geologists, and the term **till** is now preferred. However, the term is still used by engineering geologists.

Boulder pavement

A concentration of scratched (**striated**) boulders at the top of a bed of poorly sorted sediment that collectively has been planed off by overriding ice. Pavements are formed as a result of glacial **abrasion** on previously deposited **till** in a terrestrial environment, or by touch down of ice onto the floor of a lake or the sea (Figure B.22).

FIGURE B.22 Boulder pavement within the Yakataga Formation, a Miocene–Pleistocene tilted succession of glaciomarine sediment on Middleton Island, Alaska. The sediment records several grounding and erosional events.

Braid plain

A low-gradient plain influenced by seasonal rivers emanating from a glacier in a mountainous environment where sand and gravel are deposited. The river channels and intervening **bars** migrate back and forth across the plain, rendering it unsuitable for establishment of substantial vegetation (Figure B.23) (*cf.* **valley train**).

FIGURE B.23 Aerial view of the braided Kahiltna River, a few kilometres south of the terminus of Kahiltna Glacier in the Alaska Range.

Braided outwash fan (synonym: alluvial fan)

A fan-shaped feature comprising the same elements as a braid plain but originating from a point source such as a gully or at the glacier margin (Figure B.24).

FIGURE B.24 Braided outwash fan of the Murchison River, Southern Alps, New Zealand, in April 2008. The point source is Murchison Glacier in the background. The river is flowing towards the camera position. At the bottom of the photograph, the outwash fan is constrained by the lateral moraine of Tasman Glacier.

Braided river/stream

A relatively shallow river with numerous branching channels. Braided rivers typically form down-valley of a glacier. They consist of migrating channels and anastomosing geometry. Separating the channels are elevated areas of sediment called **bars**. Braided rivers carry sand, gravel and suspended sediment (silt and clay) (Figure B.25).

FIGURE B.25 Low-level aerial view of channels and bars in the braided Tasman River, New Zealand. Water is flowing from right to left and originates from glacial lake Tasman.

Brash ice

Accumulations of floating ice comprising fragments less than 2 m across, normally arising from calving glaciers or the break-up of icebergs (Figure B.26).

FIGURE B.26 Brash ice derived from tidewater Cascade Glacier in Harriman Fjord, Prince William Sound, Alaska (July 1989).

Breach-lobe moraine

A ridge of glacial debris formed where a lobe of ice breaches a lateral moraine during a glacier advance and pushes the debris beyond its former limits (Figure B.27). This ridge is typically arc-shaped and connected at each end to the older ridge.

FIGURE B.27 Breach-lobe moraine near the snout of Glacier du Miage, Mont Blanc Massif, Italy. The Little Ice Age lateral moraine was breached by the glacier, thereby creating the lobate moraine which encloses a lake with a calving ice cliff (August 2006).

Breached watershed

A short glacially eroded cross-watershed valley linking two major valleys on either side of a mountain divide (Figure B.28).

FIGURE B.28   Breached watershed at Märjela, Switzerland. During the last major glacial period, Grosser Aletschgletscher, spilt over this col into Fieschertal, a valley further east and behind the photographer. In the background, lateral moraines from the Little Ice Age are visible.

Breccia (see **ice breccia;** Figure I.15)

Brittle deformation

The failure of a material by fracturing under stress. As applied to ice, brittle failure is indicated by the formation of **crevasses** and a variety of faults (Figure B.29).

FIGURE B.29   Extensive brittle fracture in upper Fox Glacier, Southern Alps, New Zealand. Multiple sets of crevasses form as the ice converges into the steep narrow tongue (April 2008).

Bulldozing

The process of pushing of sediment in front of a glacier during an advance (Figure B.30). The sources of the debris are the immediate ice-contact sediments and any debris falling off the snout. The resulting ridges are referred to as **push moraines**.

FIGURE B.30 The advancing front of Briksdalbreen, southwest Norway, in 1996, the last time this glacier advanced. It is pushing up a ridge of debris into post-Little Ice Age birch woodland.

Bullet-shaped boulder

A boulder that is embedded in subglacial sediment (till), with a surface that is smoothed and striated on its top and stoss (up-glacier) side, and a rough plucked surface on the lee (down-glacier) side (Figure B.31). This is the boulder equivalent of a **roche moutonnée** (*cf.* **stoss-and-lee form**).

FIGURE B.31 Bullet-shaped boulder, about half a metre long, embedded within glacial sediment at Fjallsjökull, Iceland. Ice flow was from right to left.

C

Calcium carbonate precipitation

Precipitation of the mineral calcium carbonate ($CaCO_3$) beneath a glacier, especially one that flows over carbonate-rich bedrock, such as limestone. Subglacial water carrying solutes releases calcium carbonate when the water freezes to the base of the glacier forming **regelation** ice. The precipitate is found on bedrock when the ice recedes, and commonly bears lineations that reflect previous glacier movement (Figure C.1). Two mineral species of $CaCO_3$ are typically found, reflecting early and late stages in the freezing cycle, respectively sparite and micrite.

FIGURE C.1 Calcium carbonate precipitate on limestone bedrock in front of Glacier de Tsanfleuron, Switzerland. Ice flow from right to left is indicated by the parallel lineations.

Calving

The process of detachment of blocks of ice from a **glacier**, **ice sheet** or **ice shelf** into water, producing icebergs. Calving is one of the contributors to ablation, and may operate in association with frontal melting and **sublimation**. Where glaciers enter the sea or lake an ice cliff usually forms, referred to as the calving front. This can either be floating (typically for **cold** or **polythermal glaciers**) or grounded (all thermal glacier types) (Figure C.2a and b). Melting below the waterline by warm ocean waters may produce a notch that facilitates ice-cliff collapse. The rate of production of icebergs is referred to as the calving flux or calving rate, expressed as mass and time. See also **dry calving**.

C

(a)

(b)

FIGURE C.2   Calving processes. (a) An explosive calving event at the terminus of grounded polythermal glacier, Kronebreen, Svalbard. The process generates bergy bits and brash ice (July 2013). (b) Birth of a tabular iceberg from the East Antarctic Ice Sheet in Lazarev Sea (1989).

Calving glacier

The name given to a **valley glacier, outlet glacier** or **ice stream** that calves into a lake or sea. Glaciers that are grounded on the sea floor or lake bottom are typically temperate or polythermal, and the calving process is the result of collapse (sometimes with forward projection) of the cliff face or ice towers (see **calving**); this produces relatively small irregularly shaped **icebergs** and bergy bits (Figure C.2a). **Cold glaciers** and some **polythermal glaciers** tend to float, and the resulting icebergs are typically tabular (Figure C.3).

C

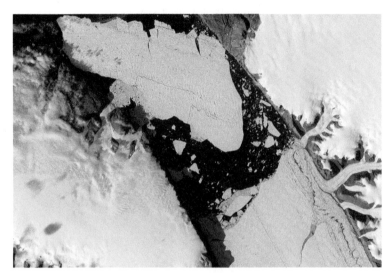

FIGURE C.3   Image of the floating tongue of polythermal Petermann Gletscher, northwest Greenland, soon after it had produced a large tabular iceberg (referred to as an ice island), acquired by NASA's Earth Observing-1 (EO-1) satellite on 16 August 2010, 11 days after calving. (Image from http://earthobservatory.nasa.gov/IOTD/view.php?id=45306)

Canal system (see **tunnel valleys**)

Candle ice

Large vertically orientated ice crystals that have grown perpendicular to the temperature gradient in glacial rivers, supraglacial ponds and other water bodies. They reach lengths of a few decimetres, and in cross-section measure a few centimetres (Figure C.4). See also **Aufeis**.

FIGURE C.4   Candle ice formed in a small pond on the western edge of George VI Ice Shelf, Antarctic Peninsula. The image shows the crystals in cross-section (November 2012).

Canyon

Exceptionally deep (>10 m) and wide (>50 m) drainage channel incised into the surface of a glacier. They develop when ice becomes less dynamic, and therefore vertically sided supraglacial channels cannot close in winter, with the result that they become larger year-by-year. On temperate glaciers they may terminate abruptly at a moulin (Figure C.5a), but on polythermal glaciers they continue over the surface to the edge or the snout (Figure C.5b). The edges of the channel tend to become less steep and straighter over time, but the active stream may retain meandering channel characteristics at the base of the canyon. In extreme cases, the canyon may cut through to the glacier bed, thereby facilitating collapse and accelerated recession.

(a)

FIGURE C.5 Canyons on valley glaciers. (a) Canyon on Vadret da Morteratsch, Switzerland, having formed over a single decade (September 2013). This canyon deepens progressively towards the photographer where it terminates abruptly as the water plunges down a moulin to the glacier bed. (b) Aerial view of two canyons on Fountain Glacier, Bylot Island, Canadian Arctic (July 2014).

(b)

Cauldron collapse

When a large cavity forms beneath a glacier or ice cap, a series of concentric fractures and crevasses form (Figure C.6). Eventually they coalesce and the ice collapses as a result. As glaciers become less dynamic during recession, cavities are formed from meltwater erosion and cauldron collapse is increasingly common. Collapse structures are typically a few tens of metres in diameter. Cauldron collapse is also induced by subglacial volcanic eruptions, such as in Iceland; these features are an order of magnitude larger.

FIGURE C.6 Cauldron collapse into a subglacial lake in the stagnating, debris-covered tongue of Unterer Grindelwaldgletscher, Switzerland (September 2009).

Cavetto form

A channel cut into a steep rock face orientated parallel to the valley side. They are the product of meltwater under high pressure and glacial abrasion. Overhanging upper lips and striations are typical.

Cavitation

Growth and collapse of bubbles in a fluid, such as subglacial meltwater, in response to pressure changes. Bubble collapse generates shock waves, which result in enhanced erosion of subglacial channels.

Chain lakes (synonym: paternoster lakes)

A chain of lakes, resembling beads on a string, in a glaciated valley. They are the product of uneven erosion by a valley glacier, forming a series of rock basins, commonly separated by low ridges of glacial sediment (recessional moraines) (Figure C.7). The term **paternoster lakes** comes from an alternative name for the Lord's Prayer of Christian tradition (derived from the Latin), and the arrangement of lakes is said to resemble rosary beads.

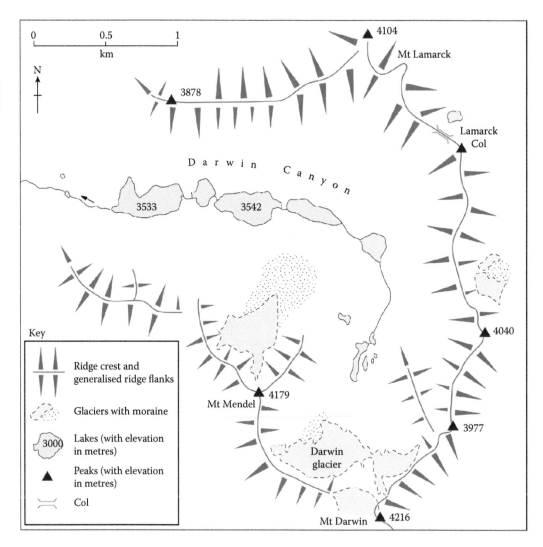

FIGURE C.7   Map of chain or Paternoster lakes in Darwin Canyon, Sierra Nevada, California, with remnant glaciers in northeast-facing cirques.

Channelled scablands

A series of anastomosing large-scale channels cut in bedrock by a huge glacial lake outburst flood. The type locality is in Washington State, USA, and this assemblage of landforms is the product of catastrophic drainage of Glacial Lake Missoula, a large ice-dammed lake ponded by the margin of the ice sheet that covered the Western Cordillera (Figure C.8a). The lake contained a maximum volume of approximately 2500 km^3, and drained several times when the ice dam failed. The flood affected an area of 40,000 km^2, and produced a range of erosional forms, notably anastomosing flood channels and gorges. The resulting depositional landforms include large-scale bars, ripples and dunes. Scablands have been recognised in other areas lying close to the edge of former ice sheets, including Antarctica (Figure C.8b).

(a)

(b)

FIGURE C.8 (a) Channelled scabland terrain in Washington State from a satellite image (NASA, false colour Landsat 7 image, 1 June 2001). The dark purple areas represent the channels, and the green and white areas the remnant plateaux in between. (b) Aerial photograph of a smaller scale region of channelled scablands, The Labyrinth, Wright Upper Glacier, Victoria Land, Antarctica. This area of channels, falls, basins, potholes and residual deposits was formed from catastrophic subglacial floods from a larger East Antarctic Ice Sheet. The edge of the current ice sheet is in the foreground (January 1996).

Chattermarks

A group of crescent-shaped friction cracks on bedrock, formed by the juddering effect of moving ice (Figure C.9). They are aligned parallel to ice flow.

FIGURE C.9 Chattermarks on Palaeogene gabbro near Loch Coruisk, Isle of Skye, Scotland, formed during the Younger Dryas glaciation of around 10,000–12,000 years BP. The 'horns' of the chattermarks point in the former up-glacier direction.

Chemical weathering

The process of breakdown and removal of bedrock by chemical processes. The rates of weathering depend on the concentration of ions in water, the area over which water is in contact with bedrock, the time that water is in contact with the bed, the availability of dissolved atmospheric gases and the reactivity of the bedrock. Chemical weathering may be particularly effective beneath a temperate glacier, and also occurs on the glacier surface and in the proglacial area. Carbonate bedrock is particularly susceptible to chemical weathering, and as a result, **calcium carbonate precipitation** occurs on bedrock surfaces and around stones (Figure C.1).

Chevron crevasses

Open linear fractures (crevasses) that trend obliquely up-glacier from the margins of a valley glacier or an ice stream (Figure C.10). They form in response to **simple shear** where there are strong lateral gradients in surface velocity.

FIGURE C.10 Chevron crevasses on the true right margin of Grosser Aletschgletscher as seen from Eggishorn, Switzerland (October 2015).

Chevron folding

A zig-zag style of folding in ice or glacial sediment. It is defined by clear compositional layering, and in ice sometimes also includes debris (Figure C.11).

FIGURE C.11   Chevron folding in sediment-rich ice at the surface of Bakaninbreen, Svalbard (August 1995). This folding is transverse to flow, and the fold hinges (crest-lines) are picked out by debris.

Chute

A vertical or near-vertical groove in bedrock, formed as a result of erosion by meltwater under high pressure at the ice/bedrock interface along the valley side (Figure C.12).

FIGURE C.12   Chute incised by glacial meltwater into Dalradian schist bedrock during the Younger Dryas Stadial. The feature is exposed at low water, adjacent to the Loch Treig reservoir, Grampian Highlands, Scotland.

Cirque (derived the French)

An armchair-shaped amphitheatre with steep sides and back wall, formed as a result of glacial erosion on a mountainside (Figure C.13). In its idealised form, a cirque contains a rock basin with a small lake or **tarn**, and a **moraine** ridge may occur on the lip. Synonyms are **corrie** in Scotland, **cwm** in Wales and **comb** in northern England.

FIGURE C.13 A group of cirques on the Glyderau mountain range viewed from Tryfan, Snowdonia National Park, North Wales. The peak in the middle is Y Garn. The lake in the centre is in the cirque named Cwm Idwal; here, Charles Darwin first recognised evidence for glaciation in Wales.

Cirque glacier

A glacier occupying a highland amphitheatre known as a **cirque** (Figure C.14a and b). A rotational component to glacier flow facilitates erosion of the cirque floor.

FIGURE C.14 Contrasting cirque glaciers. (a) A remnant cirque glacier, Teton Glacier, in the Grand Tetons National Park, Wyoming, USA, has deposited a large moraine at the lip of the cirque, which is probably still ice-cored (1988). *(Continued)*

(a)

(b)

FIGURE C.14 (Continued)   (b) A steeply graded cirque glacier on Hochfirst, Tirol, Austria. Until recently, it was connected to Gaißbergferner, the valley glacier below (September 2014).

Clast

A rock fragment or stone within a sediment or rock that is larger than the surrounding material or matrix. Clasts are a feature of glacial sediments, and range in size from granule to boulder. Analysis of clast populations is an important aspect of glacial sedimentology. Sediment attributes that are examined include: (a) fabric analysis, which is the three-dimensional orientation of the clasts (Figure C.15a); (b) clast shape and roundness (Figure C.15b); and (c) surface features such as abrasion marks, including striations, facets and crescentic gouges (Figure C.15b). Clast fabric and shape analysis can be applied to modern sediment, deposits from the last ice age and even deposits from ancient glaciations. Commonly, 50 clasts are sufficient to quantitatively define these parameters.

C

(a)

(b)

FIGURE C.15   Analysis of clasts in the Wilsonbreen Formation, at Dracoisen, Ny Friesland, Svalbard. This is a Late Precambrian (Neoproterozoic) glacial sequence implicated in the Snowball Earth theory. (a) Doug Benn measuring the orientation of clasts using a compass/clinometer. (b) A representative selection of clasts, sorted and ready for shape analysis; note that the largest clast is striated, indicating subglacial transport.

Clastic dykes

Also referred to as a **sedimentary dykes**, this feature has two origins: (a) injection or intrusion of soft sediment into a fissure extending into the beds beneath, a process which can take beneath a glacier where **hydrofracturing** is facilitated by high subglacial water pressures (Figure C.16a) and (b) as periglacial frost-wedges that become filled with sand, typically in glacial sediment (**till**) in exposed areas in front of a glacier (Figure C.16b).

(a)

FIGURE C.16 Two types of clastic dyke: (a) Clastic dykes resulting from hydrofracturing in glaciofluvial sand and gravel, Sólheimajökull, southern Iceland. (Photograph by Daniel Le Heron.) *(Continued)*

FIGURE C.16 (Continued)   (b) A probable sand-filled frost-wedge penetrating diamictite of subglacial origin in the late Precambrian (Neoproterozoic) Port Askaig Formation, Garvellach Islands, western Scotland.

Col (derived from the French)

A high-level pass formed by glacial breaching of an arête or mountain mass (Figure C.17). In the high Alps, for example, a col may be defended by steep rocky slopes.

FIGURE C.17   A col (mid-left) between Rotmoostal on the right and the valley of Gaißbergferner in front of the ice-covered mountains (left background; Tirol, Austria).

Cold (or cold-based) glacier (synonym: dry-based glacier)

A glacier in which the bulk of the ice is below the pressure melting point, and therefore frozen to the bed (Figure C.18). Although the term implies an absence of liquid water, crystal boundaries in the ice may have veins of water even at temperatures well below freezing. Water can also run off the glacier surface as a result of radiative melting, especially where debris is present.

FIGURE C.18   Wright Lower Glacier, a typical cold glacier in the Dry Valleys, Victoria Land, Antarctica. The stream flows off the glacier surface and is the source of the Onyx River, the longest in Antarctica (January 1999).

Cold ice

Ice which is below the pressure melting point, and therefore dry.

Cold-ice moraines (see **recessional moraines**)

Comb (pron. koom)

The northern English topographic name given to a **cirque**, especially in the Lake District.

Comminution

The process of creating a powder through crushing and abrasion of rock fragments as they are dragged along at the base of a glacier. The product on deposition is a **comminution till**, which has stones set within a dense fine-grained matrix.

Compound cirque

A large cirque, several kilometres across, that is wide, relative to its headwall-to-lip dimension. At the back of this amphitheatre, two or more subsidiary cirques are present, usually eroded out at a higher level than the main cirque (Figure C.19).

C

FIGURE C.19   The compound cirque of Coire Bà, bordering Rannoch Moor, Grampian Highlands, Scotland.

Compressive (or compressing) flow

The character of ice flow where a glacier is slowing down and the ice is being compressed and thickened in a longitudinal direction (Figure C.20). See also **extending flow**.

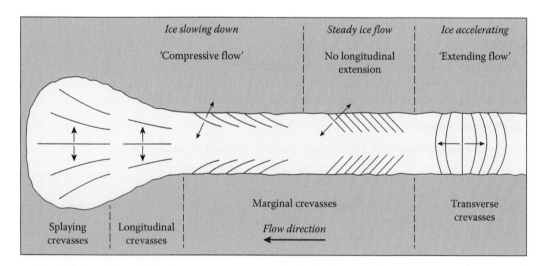

FIGURE C.20   Diagrammatic representation of compressive and extending flow in a valley glacier in plan view, with the theoretical arrangement of crevasses, based on concepts developed by John Nye in the 1950s. Compressing flow occurs where the ice decelerates, and extending flow where it accelerates. The small arrows indicate the orientation of the maximum tensile stress, and are normal to the crevasses.

Concentric crevasses

Curved open fractures that together form a ring tens of metres across on the glacier surface. They are the precursor to **cauldron collapse** and are formed as a result of the growth of a large cavity from meltwater erosion at the base of the glacier (Figure C.21).

FIGURE C.21   Concentric collapse crevasses near the true right margin of Grosser Aletschgletscher, Switzerland; telephoto view from Eggishorn. The collapse most likely was associated with a subglacial cavity caused by a stream entering from a tributary valley, a short distance up-glacier (October 2015).

Conchoidal fractures

Curved fractures on sand- and silt-sized quartz grains that are the product of glacier transport, occurring especially in the matrix of glacial sediment (**till**). They are visible only in images taken with a scanning-electron microscope.

Conduit (see **subglacial conduit**)

Congelation ice

The addition of new ice to the base of sea ice by freezing when young ice is not otherwise thickened by rafting or ridging. Congelation ice has a columnar crystal structure resulting from downward growth of crystals. When first formed, it is dark grey in appearance.

Contorted moraine

Debris on the surface of a glacier or ice sheet arranged in irregular stripes. The debris originates from the bed, and is disseminated in ice that has been contorted or folded as it impinges on an embayment. The debris may form a continuous cover on the ice surface as a result of prolonged ablation over millennia. Contorted moraine is especially common where the Antarctic Ice Sheet encounters mountain ranges that impede its flow (Figure C.22).

C

FIGURE C.22   Contorted moraines at the edge of upper Shackleton Glacier, where it enters an embayment within the rock mass of Bennett Platform. Shackleton Glacier is a major outlet of the East Antarctic Ice Sheet, flowing into the Ross Ice Shelf (January 1996).

Controlled moraine

Concentrations of glacier-surface debris that become linear mounds (**hummocky moraine**), reflecting their inheritance from structures within the parent glacier (Figure C.23). Debris is entrained from the base of the glacier by thrusting, folding and supercooling, and transported to the surface, before being released. For folding and thrusting, polythermal conditions are favoured, whereby the snout area is frozen to the bed and the mid-upper part of the glacier is wet-based. When first formed, controlled moraines have ice cores, and their morphology may reflect the orientation of the structures in the ice, but as the ice cores decay this relationship is weakened. The typical end product for a valley glacier, such as produced during the last major advance (Little Ice Age or Neoglacial), is an arcuate zone of aligned hummocks tens of metres high and 1 or 2 km wide.

FIGURE C.23   Controlled moraine forming at the snout of Midtre Lovénbreen, Svalbard. The receding margin is characterised by debris-entrainment from the base by thrusting, producing transverse ridges, and supraglacially derived sediment that has been folded parallel to medial moraines. Recently deposited controlled moraine is visible to the left of the river (August 2013).

Core-logging

Documentation of the physical properties of drilled cores of sediment or ice. Sediment cores have primarily been taken from the deep ocean, continental shelves or lakes, giving records of climatic and environmental change spanning tens of millions of years (Figure C.24). **Ice cores** have been obtained from the full thickness of the Antarctic and Greenland ice sheets, which, by providing samples of past atmospheres, have yielded detailed records of climatic change over several hundred thousand years. Ice cores have also been taken from mountain glaciers and ice caps to provide a record of climate change and atmospheric pollution on a centennial timescale (Figure I.23).

FIGURE C.24   Paul Robinson logging sediment cores obtained from McMurdo Sound during New Zealand's CIROS-1 drilling project in 1986. This is part of a 702-m-long core obtained from the sea floor that extended the then-known glacial record back to 34 million years.

Cornice

An overhanging accumulation of wind-blown snow, typically found on the edge of a cliff or ridge. Cornices are prone to collapse, and the resulting snow avalanches add to the accumulation of a glacier's mass (Figure C.25).

FIGURE C.25   Cornices at the summit of Breithorn, Wallis, Switzerland, a peak which overlooks Gornergletscher.

Corrie (from the Gaelic *coire*)

A British (Scottish) term for a **cirque** (Figure C.26).

FIGURE C.26   A large corrie named Coire Ardair on the mountain Creag Meagaidh with the enclosed tarn of Lochan a' Choire in the foreground.

Corrie with moraine

There is sometimes a close association between the erosional form of a corrie (**cirque**) and its depositional products in the form of ridge of sediment (**terminal moraine**) at the lower edge of the feature (Figure C.27). This is particularly true if the corrie has been excavated deeply enough to form a lake (**tarn**). The moraine comprises a mixture of debris eroded at the base of the glacier, and supraglacial debris from the headwall, and represents a stable position of the glacier within its self-contained basin.

FIGURE C.27   View down into Bleawater Corrie from High Street, an elongated plateau in the English Lake District. The outer edge of the corrie is marked by a prominent terminal moraine.

Crag-and-tail

A glacially eroded rocky hill with a tail of glacial sediment (**till**) formed on the lee side of the hill (Figure C.28).

FIGURE C.28 Hohentwiel, a volcanic neck near Singen, southern Germany, is part of a crag-and-tail feature. Ice flow was from right to left, and the feature was produced by ice flowing from the Alps during the Last Glacial Maximum (the Würm Ice Age).

Crater glacier

A glacier situated within or filling a volcanic crater (Figure C.29).

FIGURE C.29 Ash from eruptions in 1953 and 2007 covers the glacier filling a dormant crater on the volcano Ruapehu, North Island, New Zealand.

Creep

Permanent deformation of a material under the influence of stress. In glaciers, creep is an important component of glacier motion (see **glacier flow**). Creep of ice comprises (a) movement within crystals, involving glide along cleavage planes or along crystal defects and (b) movement between crystals accompanied by recrystallisation at crystal boundaries. The relationship between long-term creep of ice and stress is defined quantitatively by **Glen's Flow Law**. The visible manifestation of creep in glaciers includes structures such as **foliation**, **folds** and **boudinage**.

Crescentic fractures (see **chattermarks**)

Crescentic gouge

A crescent-shaped scallop, usually several centimetres across, formed as a result of bedrock fracture under moving ice (Figure C.30). Two fracture surfaces are apparent: a gently sloping surface dipping into the bedrock in the direction of ice flow, and an abrupt steep fracture facing up-glacier. The gouge is usually aligned transverse to flow.

FIGURE C.30 Crescentic gouges indicating ice movement from right to left during the last, Weichselian, glaciation on the shore of Raftsundet. The peaks of Austvågøya, Lofoten, Norway, are in the background.

Crevasse

A deep V-shaped cleft formed in the upper 'brittle' part of a glacier formed when tensile stresses exceed the tensile strength of the ice (Figure C.31a and b). This occurs when the ice is undergoing extension, such as in the accumulation area, at the margins, when it flows over a rock step or enters a water body. Crevasses are commonly hidden in whole or part by **snow bridges**, which are prone to collapse when crossing them. Crevasses in **temperate glaciers** reach a few tens of metres in depth, probably kept closed at depth by compressive stresses, but if filled with water they can propagate to greater depths. In **cold glaciers**, crevasse depths may be much greater. **Crevasse traces** indicate that in some cases, fracture may occur to the glacier bed. Crevasses provide conduits for meltwater and may become infilled with supraglacial debris. There is a wide range of geometrical arrangements of crevasses, which are featured separately in this Atlas: **arcuate, basal, chevron, concentric, *en echelon*, longitudinal, low angle, marginal, splaying** and **transverse**.

(a)

(b)

FIGURE C.31 (a) Exploring a crevasse near the snout of Vadret da Morteratsch for the Scottish television series 'Men of Rock' in May 2010. (b) Descending a rope ladder to document stratification in firn, Charles Rabot-breen, northern Norway, in 1971.

Crevasse-squeeze ridges

A series of spaced ridges made up of subglacial sediment (**till**), a few metres high and tens of metres long, that form as a result of the till being squeezed up into basal crevasses (Figure C.32a). The process is associated with glacier surges, when high stress gradients coincide with elevated basal water pressures. They are prominent in recently deglaciated terrain, notably in Iceland (Figure C.32b).

(a)

FIGURE C.32 Crevasse squeeze ridges. (a) Squeezing of saturated subglacial sediment into a basal crevasse during the surge of Comfortlessbreen, Svalbard (July 2009). *(Continued)*

(b)

FIGURE C.32 (Continued) (b) Crevasse-squeeze ridge in the proglacial area of Eyjabakkajökull, an outlet glacier on the north side of Vatnajökull, Iceland. This glacier last surged in 1972. (Photograph by David J.A. Evans, August 1995.)

Crevasse trace

A fracture trace that extends across the surface of a glacier. There are two main types: (a) a long vein of clear ice a few centimetres wide, formed as a result of fracture and recrystallisation of ice under tension without separation of the two walls; this structure commonly forms parallel to an open crevasse or extends laterally into one (Figure C.33a and b). This type of crevasse trace is analogous to a tensional

(a)

FIGURE C.33 The two main types of crevasses trace. (a) Abundant, intersecting crevasse traces of the tensional vein type in Stagnation Glacier, Bylot Island, Canadian Arctic (July 2014). *(Continued)*

C

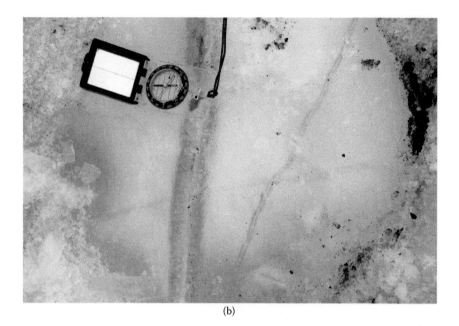

(b)

(c)

FIGURE C.33 (Continued)   (b) Detail of crevasse traces of the tensional vein type showing crystals growing towards a central suture, in Trapridge Glacier, Yukon, Canada. (c) Crevasse traces of the frozen-water type, transverse to flow in a branch of Trapridge Glacier; they were formed during a glacier surge (July 2006).

vein in deformed rock. (b) A thicker vein of clear ice resulting from the freezing of standing water in an open crevasse is also called a crevasse trace (Figure C.33c). Groups of crevasse traces may be extensively developed in a glacier, and different generations may intersect one another.

Cross-bedding/cross-stratification

The geometry of bedded sediments where sets of inclined beds or laminae are separated by original near-horizontal bounding surfaces. Cross-beds merge gradually with the lower surface but may be truncated at the top. They are formed by the lateral transport of sediment by water or wind. Water currents generate cross-beds of low-medium dip, whereas wind generates steeper cross-beds (typically 30°). Cross-bedding (centimetre to metre scale) in sand and gravel is associated with glaciofluvial transport (Figure C.34), and cross-lamination (centimetre to millimetres scale) with transport in a glaciolacustrine or glaciomarine setting.

FIGURE C.34   Large-scale cross-bedding in sand and gravel, Tonfanau, West Wales. This feature forms part of a grounding-line fan, an ice-marginal depositional landform generated by entry of a subglacial stream into a lake at the margin of the Late Glacial Irish Sea Glacier or Ice Stream.

Crushing

The process of crushing of grains and rock fragments takes place at the interface between a glacier and its bed. The particles are fractured during the shearing process and are associated with abrasion. Crushing gives a particle-size distribution that is coarser than that of **abrasion**, giving rise to non-uniform distribution of size of particles in glacial deposits (*cf.* **comminution**).

Cryoconite

Dark fine-grained sediment on the surface of a glacier. Since it absorbs more radiation than the surrounding ice, it melts downwards at a faster rate, producing **cryoconite holes** (Figure C.35a). Cryoconite provides a host for microbial activity, such as algae, which bind the sediment particles into larger granule-sized grains (Figure C.35b). The presence of microorganisms affects the carbon balance of a glacier (*cf.* **microbial processes**).

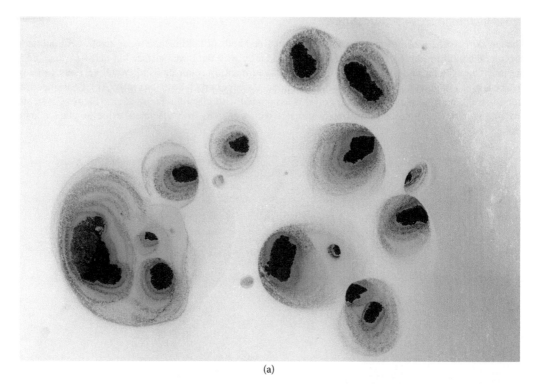

(a)

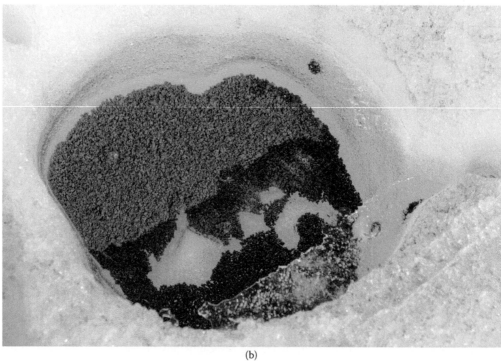

(b)

FIGURE C.35   (a) Cryoconite holes containing cryoconite on the surface of Gornergletscher, Switzerland. The holes range from 2 to 20 cm in diameter. (b) Cryoconite sediment particles bound together by microorganisms to produce a granular texture, Fountain Glacier, Bylot Island, Canadian Arctic. This hole is about 20 cm in diameter (July 2014).

Cryoconite quirk

An unusual temporary feature of cryoconite holes, whereby an arched dome of elongate radiating ice crystals forms a dome or partial dome over a cryoconite hole during sub-zero temperatures (Figure C.36).

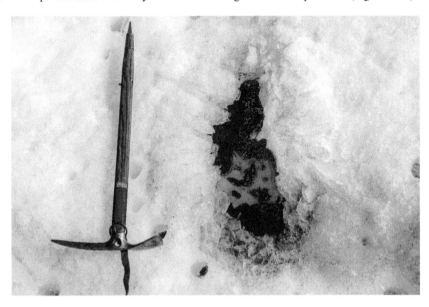

FIGURE C.36 Cryoconite quirk on Fountain Glacier, Bylot Island, Canadian Arctic. The radiating elongated crystals (frozen water) form a near-complete arched lid over a cryoconite hole (July 2014).

Cryosphere (from the Greek *kryo*, meaning 'cold')

A general term to embrace all ice-covered areas on planet Earth, including glaciers and ice sheets, sea ice, lake ice, ground ice and snow cover (Figure C.37).

FIGURE C.37 A view of Earth from space, showing the extent of the cryosphere in the Northern Hemisphere and cloud cover. NASA compilation of several images acquired on 26 May 2012 by polar-orbiting Suomi-NPP satellite.

Crystal growth (see ice crystals)

Crystal quirk

A radiating array of elongate coarse clear ice crystals, a few metres in diameter, forming cylindrical structures exposed at the surface of a glacier. A crystal quirk forms in a vertical meltwater shafts (**moulin**) that has become closed off and filled with standing water. The water freezes from the edge of the moulin towards the middle and, when the crystals meet, the crystal quirk is complete (Figure C.38).

FIGURE C.38 Large crystal quirk on the surface of Fountain Glacier, Bylot Island, Canadian Arctic. This quirk has trapped wind-blown dust and cryoconite.

Cumulative mass balance

The mass of a glacier (M) at a particular time, for example, the present (t_1) in relation to its mass at some former time (t_0), expressed as a function of time (Figure C.39). Mathematically, cumulative mass balance = $M(t_1) - M(t_0)$, where M is glacier mass, t_0 is start time and t_1 is end time.

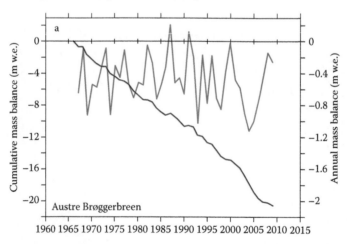

FIGURE C.39 Cumulative mass balance of a representative glacier in Svalbard, Norway: Austre Brøggerbreen. The period covered is from 1966 to 2009. Cumulative mass balance is indicated by the red line and the left-hand axis, and annual mass balance by the blue line and the right-hand axis. The zero line represents steady state; thus the glacier has experienced almost continuous decline over this period, with a cumulative loss of around 20 m of water equivalent (m w.e.). After Figure 5 in Cogley, J.G., R. Hock, L.A. Rasmussen, A.A. Arendt, A. Bauder, R.J. Braithwaite, P. Jansson, G. Kaser, M. Möller, L. Nicholson and M. Zemp, 2011, Glossary of Glacier Mass Balance and Related Terms, IHP-VII Technical Documents in Hydrology No. 86, IACS Contribution No. 2, UNESCO-IHP, Paris.

Cumulative strain (synonyms: total strain, finite strain)

The total amount of strain that a material (e.g. rock or ice) has undergone, usually in response to the prolonged application of stress. In diagrammatic form, cumulative strain is expressed in the form of strain ellipses.

Cupola hills (from the Danish *kuppelbakke*)

A group of irregular hills resulting from deformation of the bed beneath a glacier or ice sheet. They have a dome-like morphology, oval or circular in plan view, with lengths up to 15 km and heights up to 100 m. Internally, they comprise detached blocks of sediment and bedrock, commonly formed of a stacked sequence of thrust and folded strata. Glacial sediment may cover and truncate the deformation structures. Following deposition, cupola hills may be further modified into streamlined forms by the overriding ice (Figure C.40).

FIGURE C.40   Cupola hills at Mud Buttes, near Monitor, Alberta. The internal structure comprising thrust Cretaceous bedrock is well seen in this cross-section. Smoothing of the upper surface is also evident. Together, these attributes indicate ice flowing from right to left. (Photograph by David J.A. Evans.)

Curved (winding) channel

Sinuous channel cut into bedrock, usually by a subglacial stream under high pressure (Figure C.41). This is a particular form of **Nye channel**.

FIGURE C.41   Part of a curved channel incised by a subglacial stream into limestone bedrock at Glacier de Tsanfleuron, Switzerland. This channel has scalloped sides and cuts across the former ice-flow direction denoted by the striations on the smooth bedrock surface.

Cwm (pron. koom)

The Welsh term for **cirque**, or more generally for a glacial valley (Figure C.42). The term has also sometimes been used outside Wales, for example, Western Cwm on Mt Everest.

FIGURE C.42   A classic cwm or cirque in Snowdonia National Park, West Wales: Cwm Cau on Cadair Idris. A small moraine occurs on the lip of the cwm adjacent to the lake.

Cyclopel and cyclopsam

A cyclopel is a graded silt–clay couplet (pair of laminae) formed by tidal processes operating in a glacier-influenced bay or fjord. A cyclopsam is a graded sand–mud couplet, formed as for cyclopel (Figure C.43).

FIGURE C.43 Cyclopsams originally formed in the marine embayment of Engelskbukta, Svalbard, but pushed ashore by an advance of the glacier Comfortlessbreen during the Neoglacial period around 1900. These sediments are exposed within a thrust–moraine complex, and are slightly faulted and folded. Compass/clinometer for scale.

D

Dead ice (synonym: stagnant ice)

Glacier ice that has ceased to flow, usually associated with a negative mass balance when mass transfer of ice is reduced (Figure D.1). Dead ice is typically covered by surface debris, which causes the ice to melt unevenly. Sometimes dead ice becomes detached from the active glacier (Figure D.1), and over time vegetation may be established on the ice surface. Alternatively dead ice may be buried by glaciofluvial sediment, which, when it slowly melts, gives rise to a pitted surface (see also Figure S.45).

FIGURE D.1 Debris-covered dead ice associated with the rapidly receding Triftgletscher, Switzerland (August 2002).

Débâcle (from the French *débâcle*)

A glacial lake outburst flood (Figure D.2), also known as an aluvión.

FIGURE D.2   Debris from the débâcle that was caused by the rupture of a water pocket within a small glacier on Tete Rousse, Mont Blanc massif, French Alps, which occurred on 11/12 July 1892. In the background are some houses in the village of Bionnay that remained undamaged. The bodies of 175 victims were recovered, but the total death toll is unknown. (Photograph by Ch. Kuss, 14 July 1892; http://arve.randonnees. free.fr/pages/catastrophe_aux_thermes_du_fayet.html)

Debris apron

A wedge of debris that accumulates in front of an ice cliff of **cold** or **polythermal glaciers**. The debris originates from englacial layering and is originally derived from the base of the glacier. It is commonly mixed with fallen blocks of ice to a greater (Figure D.3a) or lesser (Figure D.3b) extent. The latter case may involve internal deformation and the surface may ablate to give a smooth ramp.

(a)

FIGURE D.3   Two types of debris apron accumulation. (a) Irregular wedge of debris and some fallen ice blocks derived from low-angle thrust-faults in the advancing polythermal Thompson Glacier, Axel Heiberg Island, Canadian Arctic. Note the person for scale to the left of the base of the waterfall (Summer 1975).
(Continued)

(b)

FIGURE D.3 (Continued)   (b) Smooth ramp comprising dry-calved ice and basally derived debris, at the near-stationary margin of the cold-based Suess Glacier, Victoria Land, Antarctica (February 1999).

Debris (dirt) cone

A thin veneer of debris draping a cone of ice up to several metres in height (Figure D.4). Dirt cones begin forming when surface sediment is concentrated, for example, by surface streams, and as the debris protects the ice beneath from **ablation**, a cone begins to grow. The sides of the cone stabilise at an angle of around 35–50°.

FIGURE D.4   Debris cone, comprising sandy gravel, originally sorted by a supraglacial stream, on Unteraargletscher, Switzerland (Summer 2004).

Debris entrainment

The processes whereby debris is incorporated into the body of a **glacier** or **ice sheet**. Several modes of entrainment occur (Figure D.5). From the surface, the processes are ingestion of **supraglacial debris** by **folding**, and within **crevasses**, **supraglacial stream channels** and **moulins**. At the base of a glacier, the main processes are incorporation in the **basal debris** layer by **regelation**, **thrust-faulting** and **folding**.

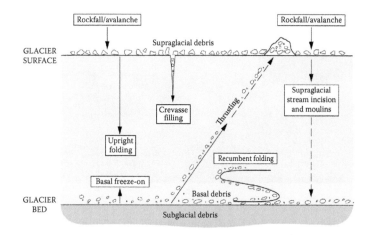

FIGURE D.5 Schematic representation of the principal processes of entrainment in a glacier. A vertical longitudinal cross-section is shown, with ice movement from left to right. (Adapted from Hambrey, M.J., et al., *Journal of Glaciology*, 45, 69–86, 1999. With permission from the International Glaciological Society.)

Debris flow

There are two types of debris flow in glacial environments. On land, a type of gravity flow involves the movement of unconsolidated material downslope as a slurry. Total disaggregation of the material takes place, and some fine material is carried away in suspension in water. This type of debris flow commonly consists of poorly sorted, saturated glacial sediment (**till**) in proximity to, or overlying, a melting ice margin, in which case it is referred to as a **glacigenic sediment flow** (Figure D.6). **Subaqueous debris flows** occur where large volumes of debris are dumped into the sea or a lake at the calving terminus of a glacier. In some cases, movement induces a degree of sorting and results in coarsely graded beds of sediment. Some of the clay and silt fraction may escape from the flow and form a turbidity current, which produces well-graded beds.

FIGURE D.6 Debris flow of subglacial debris (till), comprising fine sediment and stones up to boulder-size at the margin of Trapridge Glacier, Yukon (July 2006). This flow is occurring on top of melting dead ice from an advance that predates the deposition of subglacial sediment. The debris flow enters a stream flowing off the glacier and is reworked.

Debris layering

The layering of debris exposed at high level in a mountain glacier. The debris originates from rockfall or may be mixed with ice and snow avalanches. It contributes to the stratified nature of the ice and is released at the surface (Figure D.7). This type of layering is common in debris-mantled glaciers. Debris layering is also an integral part of **basal ice**.

D

FIGURE D.7 Debris layering derived from avalanches in Lhotse Glacier, Khumbu Himal, Nepal. Release of debris by melting streaks the ice cliff with sediment, and some collects in the adjacent pond (May 2003).

Debris-mantled (debris-covered) glacier

A mountain glacier whose ablation area is largely covered by rock debris (Figure D.8). The debris is largely derived from rockfall, but a component of basal debris may also reach the surface by deformational processes. In the accumulation area, rock debris is mixed with snow and avalanche material. As this moves into the ablation area, it melts out, adding to the concentration of debris at the surface.

FIGURE D.8 The debris-mantled surface of Khumbu Glacier, Mt Everest Region, Nepal, with the peak of Pumori in the background. The debris varies considerably in thickness as the surface ablates unevenly, but on average is 1–2 m thick. Ponds are common on such glaciers and eventually coalesce to form unstable lakes (April 2003).

Debris septa

Layers of debris within a glacier that crop out at the surface as lines of debris, usually forming part of a **medial moraine** (Figure D.9). The debris originates from various sources, such as from the bed, or from supraglacial sources and modified into lines by glacier flow. **Folding** may also be implicated in the formation of septa, with the debris being aligned parallel to the fold axis.

D

FIGURE D.9   Debris septa emerging near the snout of Feiringbreen, Kongsfjorden, Svalbard. The lines of debris are the surface expression of folded supraglacial debris, and trend parallel to fold axis, which in turn are parallel to ice flow (July 2013).

Debuttressing

The process of large-scale rock failure when valley walls lose the support (or buttressing effect) of a glacier during or following recession (Figure D.10) (*cf.* **rock-slope failure** and **pressure release**).

FIGURE D.10   A recent collapse of a wall of limestone following rapid thinning of Unterer Grindelwaldgletscher. The glacier tongue still lies beneath this rockfall (September 2009).

Deformable bed

The soft bed of sediment beneath a glacier or ice stream that deforms as the ice flows over it. The sediment is till, containing clay-to-boulder sized material, which behaves like slurry when it is saturated with water and under high pressure (Figure D.11). Deformation of the bed is one of the three main components of **glacier flow**.

FIGURE D.11 The terminus of the partly tidewater glacier Conwaybreen in northwestern Spitsbergen, Svalbard. The deformable bed is in the foreground. It consists of a layer of poorly sorted sediment (till), the muddy matrix facilitating deformation when saturated (July 2009).

De Geer moraines

Named after a Swedish geologist (1858–1943), these are a group of moraines formed subglacially, transverse to flow, beneath a glacier terminating in a lake or the sea. They have the form of discrete narrow ridges that are well-spaced (Figure D.12).

FIGURE D.12 De Geer moraines exposed at low tide in Ymerbukta, Isfjord, Spitsbergen, Svalbard. (Photograph by Julian A. Dowdeswell, September 2013.)

Deglaciation

The process of transition on a global scale from full glacial conditions during an ice age, to a warmer interglacial period. This change is manifested in sea level rise as a result of shrinkage in the volume of continental ice. At the local scale, deglaciation refers to the recession of an ice mass on a shorter timescale, for example, since the Little Ice Age, revealing the former bed of the glacier and allowing regeneration of vegetation (Figure D.13).

FIGURE D.13 Ongoing deglaciation since the Little Ice Age at Athabasca Glacier, Canadian Rocky Mountains. The glacier has receded 1.5 km since the late 19th century, when the glacier reached up to the lateral moraines on the left of this photograph. Low shrubs are beginning to regenerate on the deglaciated foreground. It is expected that this glacier, and the Columbia Icefield which supports it, will have largely disappeared by the end of the century. Athabasca Glacier is easily accessible from the Icefield Parkway and marker posts indicate the position of the glacier snout at regular time intervals. Tour buses run onto the glacier along the lateral moraine (June 2006).

Delta–fan complex

A prograding (outward-building) sequence of glaciomarine sediment, gravity-flow material and basal glacial debris, formed at the grounding-line of an ice shelf or ice stream on the continental shelf (*cf.* **grounding-line fan**).

Delta moraine (synonym: kame delta)

A delta complex formed in an ice-frontal position, commonly below water level, where subglacial streams enter a proglacial lake or the sea. Such features are composed of a mixture of glaciofluvial and glacial debris (*cf.* **grounding-line fan**).

Dendritic drainage (synonym: arborescent drainage)

A network of surface streams on a glacier that resembles the branches of a tree (Figure D.14). Small streams ('branches') combine down-glacier to form a progressively larger 'trunk' stream.

FIGURE D.14 Dendritic drainage system on Vadret da Tschierva, a temperate glacier in Kanton Graubünden, Switzerland, September 2009. The width of the drainage system in the middle of this image is about 20 m across.

Desiccation

The process of cracking of sediment, usually **mud** in a glacial environment, as it dries out. The cracks are open by a few millimetres and have a polygonal arrangement (Figure D.15). Mudflats occur where suspended sediment has been deposited from glacial streams, or where the mud fraction of **debris flows** separates out and collects in hollows.

FIGURE D.15 An area of mud, winnowed out from basal glacial sediment during a debris flow on recently exposed ground on Ossian Sarsfjellet, northwestern Spitsbergen, Svalbard. As this mud has dried, desiccation cracks have formed in a polygonal arrangement. Polygon width is about 10–20 cm.

Diagenesis

The changes that occur in a sedimentary sequence following deposition. In glaciers, this is represented by the transformation of snow to **firn** and then ice. In sediment, the process involves compaction (lithification) into rock, possibly in association with chemical changes such as cementation.

Diamict apron

A prograding sequence of lateral continuity comprising diamicton and to a lesser extent sand, gravel and rhythmite. Its morphology and sedimentary attributes are defined by seismic surveys and drill-core analysis, respectively. This is a non-genetic term, intended to replace the term **till delta**, which is imprecisely defined sedimentologically. A typical interpretation, however, would be a **grounding-line wedge** of sediment, comprising glacial sediment and reworked sediment, formed just seawards of an ice stream or glacier **grounding-line/zone** at a steepening break in slope on the sea floor. If formed at the mouth of a trough that has been glacially eroded to below sea level, the more specific term **trough-mouth fan** is used.

Diamicton/diamictite/diamict

A non-sorted terrigenous sediment containing a wide range of particle sizes, commonly from clay to boulders. **Diamicton** is the unconsolidated variety, and **diamictite** the lithified variety (Figure D.16). **Diamict** is a term that covers both. Texturally, there is no universally agreed scheme for classification of poorly sorted sediments, including diamicton(-ite), so usages vary considerably. However, one scheme based on the relative proportions of sand, gravel and mud, which has been widely used, is illustrated in Figure D.17.

FIGURE D.16 Diamictite in the early Proterozoic (c. 2000–2500 million years) Gowganda Formation at Whitefish Falls, Ontario, Canada. This rock is interpreted as a glaciomarine sediment belonging to an ice age that spanned several continents.

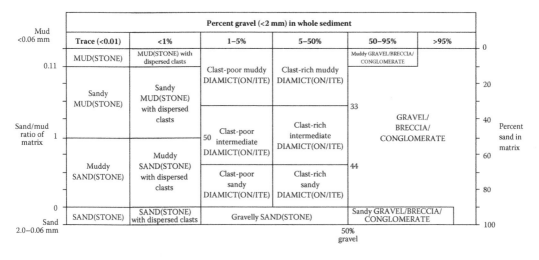

	Percent gravel (<2 mm) in whole sediment						
Mud <0.06 mm	Trace (<0.01)	<1%	1–5%	5–50%	50–95%	>95%	
0.11	MUD(STONE)	MUD(STONE) with dispersed clasts	Clast-poor muddy DIAMICT(ON/ITE)	Clast-rich muddy DIAMICT(ON/ITE)	Muddy GRAVEL/BRECCIA/ CONGLOMERATE		0
	Sandy MUD(STONE)	Sandy MUD(STONE) with dispersed clasts					20
Sand/mud ratio of matrix 1			Clast-poor intermediate DIAMICT(ON/ITE) 50	Clast-rich intermediate DIAMICT(ON/ITE)	33 GRAVEL/ BRECCIA/ CONGLOMERATE		40
	Muddy SAND(STONE)	Muddy SAND(STONE) with dispersed clasts	Clast-poor sandy DIAMICT(ON/ITE)	Clast-rich sandy DIAMICT(ON/ITE)	44		60 / 80
0 Sand 2.0–0.06 mm	SAND(STONE)	SAND(STONE) with dispersed clasts	Gravelly SAND(STONE)		Sandy GRAVEL/BRECCIA/ CONGLOMERATE		100

Percent sand in matrix

50% gravel

FIGURE D.17 Classification of poorly sorted sediments, including diamict, based on textural classification in the field. First, the proportion of gravel is investigated, then the ratio between sand and mud in the matrix. (From Hambrey, M. J. and Glasser, N. F., Glacial sediments: processes, environments and facies, In Middleton, G. V. (ed.) *Encyclopedia of Sediments and Sedimentary Rocks*, Kluwer, Dordrecht, 316–331, 2003.)

Differential weathering (of glaciers)

Unequal reaction to different crystallographic types of ice (see **facies [glacier ice]**) or ice structure under the same weather conditions. Coarse bubbly ice is darker than adjacent coarse bubbly ice, so ablated more quickly. Fine-grained ice is softer than these types of ice when weathered, and may melt more quickly. Debris content also influences the rate of weathering, with debris-bearing ice melting more quickly until the surface cover increases sufficiently to slow down melting. Differential weathering is especially noticeable at the glacier surface in (a) steeply dipping **longitudinal foliation** (Figure D.18), which produces a furrowed surface, and (b) where gently inclined dirty layers in stratification or along thrust-faults intersect the surface, producing small overhangs.

FIGURE D.18 Differential weathering of the surface of Vadrec del Forno, Switzerland. The furrows reflect the more rapid melting of coarse clear ice, compared with the ridge of coarse bubbly ice. Prolonged exposure to solar radiation enhances this effect (September 2011).

Diffluence

The state of glacier flow whereby a glacier divides and overflows into adjacent valleys *via* low breaches or cols (Figure D.19). The name given to this type of feature is **transection glacier**.

FIGURE   D.19 Diffluence of a rapidly receding outlet from the Southern Patagonian Icefield, Chile. The tongue at top right is Glaciar Occidental, and the middle and lower tongues are known collectively as Glaciar Greve (December 2004).

Dilation

The expansion in volume of a material after release of stress. It commonly occurs in bedrock following ice-removal, and leads to fracturing (see also **exfoliation,** Figure E.9). It also occurs in sediment, especially if saturated.

Discharge (also referred to as **Runoff**)

The rate of flow of water or ice through a vertical section perpendicular to the flow direction. Discharge from glaciers may be monitored for assessment of water resources, especially where used for hydroelectric power generation (Figure D.20). Discharge from glaciers varies on a daily basis (**diurnally**) and seasonally. The unit of measurement is usually cubic metres per second (cumecs).

FIGURE   D.20   High discharge   from   Vernagtferner, Ötztaler Alpen, Austria, passing a gauging station. Note the large amount of suspended sediment. (Photograph by Markus Weber.)

Dirt cone (see **debris cone**)

Distributed drainage system

The state of drainage beneath a glacier at times of high water discharge, when basal water pressures are high, leading to separation of ice from the bed over wide areas. This allows the ice to flow more rapidly and is believed to be particularly associated with the early part of the melt-season.

Diurnal fluctuation/variation

Variations taking place on a day–night cycle, commonly referring to changes in discharge from a glacier over a 24-h period (Figure D.21).

FIGURE D.21 Diurnal flow of the outlet river from Vadret da Morteratsch, Switzerland. The left-hand photograph was taken at 9.33 a.m., local time, and the right-hand photograph at 1.13 p.m. on the same day, 3 September 2013. Note how at peak flow most of the boulders on the river-bed and the gravel bar are covered by water, and that the suspended sediment load is higher.

Doline

An ice-collapse feature that forms in the middle of an ice shelf, as a result of melting from beneath. These oval-shaped depressions may reach several kilometres across and usually have a floor of sea ice (Figure D.22).

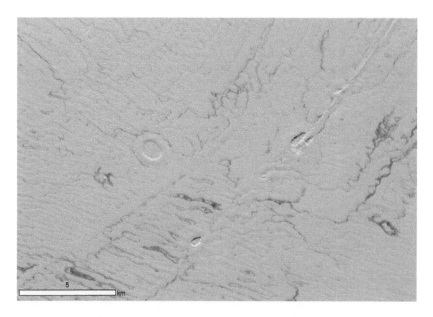

FIGURE D.22 Dolines (oval depressions) in George VI Ice Shelf surrounded by a strong network of structurally controlled drainage channels. Extract from an ASTER image, prepared by Tom Holt.

Dome

A smooth, rounded boss of glacially abraded bedrock, commonly exceeding a hundred metres in diameter and height (Figure D.23).

FIGURE D.23 Domes of granite at Tenaya Lake, Yosemite National Park, California, USA. They are glacially abraded on all sides.

Downwasting

Thinning of a glacier as a result of ablation (Figure D.24). Most glaciers in the world are rapidly downwasting, as well as **receding** at the time of writing (2015), with annual losses amounting from a few to several metres a year.

FIGURE D.24 The Khumbu Glacier, Mt Everest region, Nepal, reveals considerable down-wasting, but little frontal retreat, since the Little Ice Age of the mid-19th century. This former level is marked by the bare moraine walls, which reach elevations of over 100 m above the present-day surface (May 2003).

Drift

Formerly applied to deposits that were formed from floating icebergs, this 19th-century term was subsequently used to describe all unconsolidated deposits associated with glaciers and glacial meltwater. The term is still widely used, especially for mapping of Quaternary deposits. Additionally, other unconsolidated deposits are sometimes included under the term 'drift' (as in the 'Drift geology' maps of the British Geological Survey).

Dropstone

A relatively large stone (**clast**) that falls through the water column into soft sediment, disrupting the bedding or laminae. Draping of sediment over the top of the clast subsequently occurs. In a glacial context, dropstones are released from icebergs (Figure D.25).

FIGURE D.25 Dropstone boulder in stratified and laminated sediments of the Sirius Group, a Neogene glaciolacustrine deposit at Bennett platform, Shackleton Glacier, Antarctica (February 1996).

Drumlin (from the Gaelic word *druim*)

A smooth hillock, often occurring in groups or swarms (Figure D.26a and b), composed of glacial debris (especially **till**, but also **mud**, **sand** and **gravel**), and sometimes having a bedrock core. Drumlins are typically a few hundred to a few kilometres in length and up to 50 m in height. The ratio of length-to-width varies from 2:1 to 7:1. A drumlin commonly has a steeper up-valley slope (stoss side), and resembles a half-buried egg, hence the term **basket-of-eggs topography**. Although the precise mode of formation of drumlins is uncertain, there is a consensus that they are formed by a combination of erosion and deposition beneath an actively flowing **glacier** or **ice stream**, and become elongated parallel to the former ice-flow direction. A less widely accepted theory of drumlin formation is that they are the result of meltwater erosion and deposition during large floods beneath the ice. Drumlins belong to a family of **subglacial bedforms** that include **drumlinoid ridges**, **flutes**, **mega-scale glacial lineations** and **Rogen moraines**.

(a)

(b)

FIGURE D.26 (a) Ground view of a drumlin field between Zürichsee and Zugersee, Switzerland. (b) Aerial view of a single drumlin with road along the crest near Grüningen, Switzerland. The ice-flow direction was towards the camera. In both cases, the drumlins formed during the late Würm glaciation.

Drumlinoid ridges (or drumlinised ground moraine)

Elongate, strongly linear ridges, intermediate in morphology between **drumlins** and **fluted moraine**. On a large scale, these features are referred to as **mega-scale glacial lineations**.

Dry-based glacier (synonym: cold glacier)

Dry calving

Ice discharge from an ice margin with cliffs onto land, usually in discrete blocks. Heaps of calved ice accumulate at the base of cliffs, where they melt or are removed by rivers). Dry calving is associated mainly with **cold** (Figure D.27a) and **polythermal glaciers** (Figure D.27b).

(a)

(b)

FIGURE D.27   Dry calving at the margins of cold and polythermal glaciers respectively. (a) Aerial view of the snout of Rhone Glacier, Victoria Land, Antarctica, where cliff collapse is induced by slowly advancing ice and the development of overhangs (January 2009). (b) Fountain Glacier, Bylot Island, Canadian Arctic, where the ice cliff has collapsed as a result of undercutting by an ice-marginal to subglacial stream (July 2014).

Dump moraine

Mounds of debris that formerly rested on the surface of a **glacier**. They have been 'dumped' as the underlying ice melted, and may include both surface and basal debris (Figure D.28). Small dump moraines may form at an active ice margin that remains stationary during the winter. Larger dump moraines up to several metres high form if the terminus is stationary for several years. If there is an advance of ice into the dumped material, **push moraines** are formed. Dump moraines may expand over time with advancing or stationary ice into large **terminal** or **lateral moraine** complexes.

FIGURE D.28 Dump moraine that forms part of the end moraine complex of Tasman Glacier, New Zealand (January 2011).

Dust

Fine-grained mineralogical or biogenic material disseminated over a glacier surface. Texturally, it is normally silt- or sand-grade. Dust is derived from the surrounding hillsides or transported high in the atmosphere from distant deserts. As glaciers ablate, dust is concentrated and several years' accumulation tend to give the glacier a greyish or sandy-coloured hue (Figure D.29). This decreases the albedo, resulting in faster melting.

FIGURE D.29 Saharan dust on the upper reaches of Glacier d'Argentière, Mont Blanc Massif, France. The orange tinge is particularly noticeable where it contrasts against the white avalanche cones at the base of the headwall (August 2008).

Dry-snow zone

The region of a **glacier**, usually at high elevation, where there is neither surface melting nor rainfall. The zone below is the **percolation zone**, and the line separating the two is the **dry-snow line**.

Dynamic thinning

The reduction of glacier thickness, in excess of ablation, that results when more mass flows out of the area of thinning than flows in. This process characteristically affects tidewater or floating glaciers, and is usually reflected in enhanced calving at the terminus (Figure D.30). Dynamic thinning also takes place at the scale of an ice sheet. It occurs in Greenland because increased surface meltwater is reaching the bed, allowing the ice to accelerate. In the case of the West Antarctic Ice Sheet, thinning of ice shelves is the result of both oceanic and atmospheric warming; the result being massive calving events.

D

FIGURE D.30   Dynamic thinning of Pine Island Glacier, a major outlet of the West Antarctic Ice Sheet, is resulting in acceleration of the ice, and the loss of over 100 gigatonnes per year from basal melting and half as much again from calving. This thinning and acceleration leads to weakening of the floating ice mass and increasing propensity for calving. In this ASTER satellite image, taken 13 November 2013, a large crack, 30 km long is evident. The resulting iceberg that calved in November 2014 covered 720 km². (Image from NASA pineisland_ast_2011317.)

E

Earth pyramids (from the French *Demoiselles*)

Pinnacles and sharp pyramid-like forms comprising basal glacial sediment (**till**) situated along the sides of alpine valleys. They form where rainwater erodes a sheet of till unevenly. Large boulders commonly cap the pyramids; these are responsible for the localised protection of the till from erosion (Figure E.1).

FIGURE E.1 Earth pyramids in the Ritten area, high above the city of Bozen (Bolzano) in the Italian Alps. These unusually pink and accessible pyramids are a famous tourist attraction. They are located about 700 m above the floor of the Eisack Valley. The sheet of glacial sediment from which they are derived was deposited by the largest glacier system of the Late Glacial Maximum, Etsch Glacier, on the southern slope of the Alps.

Eccentricity

Change of shape of the Earth's orbit from almost circular to elliptical and back. These changes take place on various cycles, of which the best known in the Quaternary ice age record is between 96 and 136 ka (thousand years). There is also a 413 ka eccentricity cycle, and other cycles may exist in the geological record. Eccentricity is one of the three astronomical variables that affect the amount of solar radiation received at the Earth's surface, and is a component of the Astronomical Theory of Climate Change.

Ecological succession

The sequential development of vegetation (and associated animals) on a glacier **forefield** after a glacier has receded (Figure E.2).

E

(a)

(b)

(c)

FIGURE E.2   Three images of the proglacial area of Vadret da Morteratsch, Switzerland, taken from exactly the same position in 1985, 2002 and 2015. Note the post was replaced in 2015. The glacier has receded substantially and the sparse shrubs and alpine plants have been progressively supplemented by larches. Later on stone pines will follow.

En echelon crevasses (from the French *échelon* meaning rungs of a ladder)

Sets of short open fractures or crevasses that form in glaciers in response to **simple shear** and rotation, such as near the margins of a valley glacier (Figure E.3). During rotation, the crevasses may attain sigmoidal shapes before closing up.

FIGURE E.3 Several sets of en echelon crevasses, cross-cutting medial moraines, near the east margin of Vadret Pers, viewed from Diavolezza, Switzerland. Ice flow is from to top left to lower right (September 2009).

End moraine (synonym: terminal moraine)

The outermost ridge of glacial debris pushed up by a glacier, and marking the limit of its advance. A wide range of processes may be involved in its formation, including dumping, pushing and thrusting. The composition is highly variable and includes sediment directly deposited from the glacier, as well as glaciofluvial, glaciolacustrine and glaciomarine sediment. In areas of permafrost, stresses generated by a glacier may propagate through the frozen ground, creating thrust ridges a few tens or a hundred metres beyond the actual ice margin (Figure E.4).

FIGURE E.4 A composite end moraine at the terminus of both land- and sea-terminating Comfortlessbreen, northwestern Spitsbergen, Svalbard, in August 1992. The complex of ridges was generated by thrusting both at the ice margin and in the permafrost ahead of it. The moraine contains glacial, glaciofluvial and glaciomarine sediment. The greater part of the ridge dates from the Neoglacial period (around 1900), but some elements are considerably older. This glacier surged between 2005 and 2009, but did not reach the limit of this moraine.

Energy balance

A relationship that describes the change in the amount of energy stored within a defined volume of ice. This change arises from the exchange of heat or radiation, which leads to changes in temperature and phase, such as melting, freezing and sublimation. Thus energy balance is strongly connected with **mass balance**. The surface energy balance of a glacier is measured using an array of meteorological instruments that run all year-round (Figure E.5). The internal and basal energy balances of a glacier are difficult to measure but can be calculated using an energy-balance model.

E

FIGURE E.5   An automatic weather station has been operated by Utrecht University (Hans Oerlemans) since 1995 on Vadret da Morteratsch, a temperate glacier in Switzerland. The meteorogical variables are air temperature, pressure, wind speed and direction, incoming and outgoing radiation (both shortwave and long-wave), humidity, snow accumulation and ice melt. From the resulting data, surface energy balances have been obtained (September 2013).

Englacial channel/conduit

A tunnel enclosed by **glacier ice**, formed by surface meltwater penetrating below a glacier surface *via* channels and vertical shafts (**moulins**). Conduits are commonly approximately spherical in cross-section (Figure E.6a), but as a stream cuts downwards it becomes elongated about a vertical axis. Conduits are prone to closure by ice **creep** in actively deforming glaciers outside the melt-season, but in inactive ice may grow into large cavities. **Englacial stream** is the term given to the flow of water in such a channel. Whereas in **temperate glaciers** englacial streams reach the bed quickly, in **polythermal glaciers** those may take long routes and emerge in elevated positions in an ice face. Some streams may connect with the basal ice zone and thus carry sediment (Figure E.6b).

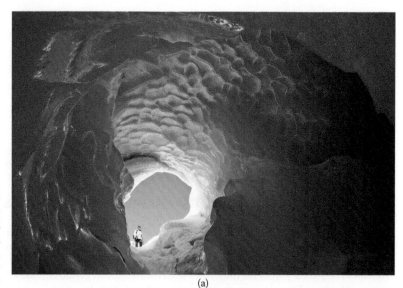

FIGURE E.6   Englacial channels in valley glaciers. (a) An abandoned englacial channel in Vadret da Morteratsch, Switzerland (February 2009).   *(Continued)*

(a)

(b)

FIGURE E.6 (Continued) (b) An englacial stream that connects with the bed, carrying a high volume of sediment, emerging from a cliff face at polythermal Crusoe Glacier, Axel Heiberg Island, Canadian Arctic (July 2008).

Englacial debris

Debris dispersed throughout the interior of a **glacier**. It originates either in surface debris that is buried in the accumulation area or falls into **crevasses**, or in **basal debris** that is raised by basal freeze-on, thrusting or folding (Figure E.7). See also **debris layering**.

FIGURE E.7 Folded englacial debris layer originally derived from the bed, exposed in the left-hand margin of Taylor Glacier, Victoria Land, Antarctica (January 2001).

Eolian sediment (North American spelling for **Aeolian sediment**)

Epi-shelf lake

A body of water enclosed by an **ice shelf** but with a connection to the open sea beneath the ice shelf. The surface is commonly frozen and subject to deformation by flow of the ice shelf and tidal changes (Figure E.8).

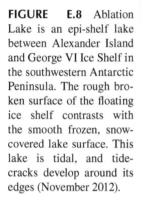

FIGURE E.8 Ablation Lake is an epi-shelf lake between Alexander Island and George VI Ice Shelf in the southwestern Antarctic Peninsula. The rough broken surface of the floating ice shelf contrasts with the smooth frozen, snow-covered lake surface. This lake is tidal, and tide-cracks develop around its edges (November 2012).

Equilibrium line/zone

The line or zone on a glacier surface where a year's **ablation** balances a year's **accumulation** (*cf.* **firn line**). It is determined at the end of the ablation season. It is sharply defined on temperate glaciers, but on polythermal glaciers it tends to be diffuse, and commonly occurs at the boundary between refrozen surface meltwater (**superimposed ice**) and glacier ice (Figure E.9a and b).

(a)

FIGURE E.9 (a) Equilibrium line on Vadret Pers, Kanton Graubünden, Switzerland (August 2015). This marked by the boundary between the white snow and brownish firn. The firn line in this case is the lower limit of firn, where it overlies blue glacier ice.   *(Continued)*

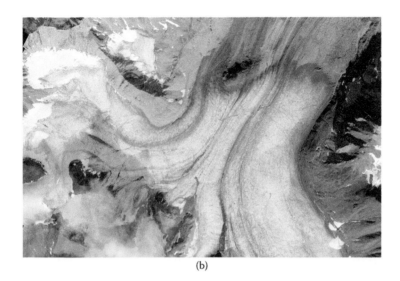

(b)

FIGURE E.9 (Continued) (b) Vertical aerial image of Austre Brøggerbreen, a composite valley glacier in northwestern Spitsbergen, Svalbard. This glacier has been subject to strongly negative mass balance since records began in the 1960s. The strong net ablation in this mid-summer image has shrunk the accumulation zone drastically. Thus the equilibrium line for this melt-season is near the upper reaches of the glacier; this is indicated by the line between white snow and greyish ice on the far left. (Image from the Natural Environment Research Council, UK, taken on 25 July 2004.)

Erratic

A boulder (Figure E.10a) or large block of bedrock that is being, or has been, transported away from its source by a glacier. Commonly, the term is limited to a block that has a different lithology from that upon which it is resting (Figure E.10b). Large erratics may sometimes be perched on smaller cobbles or boulders (see also **glacially transported boulder**).

(a)

FIGURE E.10 (a) A one-metre-long erratic of dark grey gneiss, perched on three cobbles that rest on top of pale grey gneiss, west coast of Isle of Harris, Scotland. *(Continued)*

(b)

FIGURE E.10 (Continued) (b) A pair of dark grey Silurian turbidite erratics resting on whitish Carboniferous Limestone, at Norber near Austwick, Yorkshire Dales National Park, England. The long axes of these to boulders are about 3 m.

Esker (from the Irish *eiscir*)

A long, commonly sinuous ridge of sand and gravel, deposited by a stream in a glacier ice-bounded **subglacial channel**. Eskers formed beneath former **ice sheets** reach several hundred kilometres in length and 50 m in height (in Canada). Eskers associated with valley glaciers are typically a few tens or hundred metres long and several metres high (Figure E.11a). The sand and gravel is well sorted and commonly cross-stratified (Figure E.11b).

(a)

FIGURE E.11 Overview and detail of an esker in front of Comfortlessbreen, a partly tidewater glacier in north-west Svalbard. (a) The sinuous nature of the small, well-formed esker.   (*Continued*)

(b)

FIGURE E.11 (Continued)   (b) the sand and gravel texture with hints of cross-bedding. Both photographs were taken in July 1992.

Eustacy (see glacio-eustacy)

Expanded foot glacier

A glacier that spreads out laterally from its confining valley walls into a lowland area. It represents the early stages in the development of the fully lobe-shaped form known as a **piedmont glacier**.

Exfoliation

The process of removal of sheets of bedrock along joints parallel to the rock surface. The release of pressure following ice-loading creates the joints (Figure E.12).

FIGURE E.12 Exfoliation of granite resulting from ice-sheet unloading, Tenaya Lake, Yosemite National Park, California.

Extending flow

The character of ice flow where a glacier is accelerating, and the ice is being stretched and thinned in a longitudinal direction (Figure E.13; see also Figure C.20).

FIGURE E.13 Extending flow indicated by abundant transverse crevasses in this composite cirque glacier that reaches the coast of Palmer Land, near the northern edge of George VI Ice Shelf (in the background), Antarctica (November 2012).

F

Fabric (of ice and till)

The texture and arrangement of all the particles and crystallographic attributes in geological materials. In glacier ice, fabric refers to ice crystals and entrained sediment, especially crystal orientation. In glacial sediment, fabric commonly applies to the orientation of stones (**clasts**), which may be a good indication of mode of deposition and ice-flow direction (Figure F.1).

FIGURE F.1 Preferred stone (clast) orientation fabric in glacial sediment: the Neogene Sirius Group, Roberts Massif, Shackleton Glacier, Antarctica. The clasts are aligned parallel to the hammer. This sediment is interpreted as a basal till.

Facet

A smooth flat surface with rounded edges on a stone (**clast**). The surface is the product of glacial **abrasion**, and **striations** may be present on suitable fine-grained lithologies (Figure F.2).

FIGURE F.2 Facetted clast with striations on a limestone boulder, close to the margin of Austre Lovénbreen, northwestern Svalbard.

F

Facies (basal ice)

The types of ice and ice/debris mixtures that are the result of the distinctive range of physical and chemical processes operating at or near the bed of a glacier or ice sheet. Facies include various types of water ice, **glacier ice** and debris-bearing ice and sediment, which have been deformed to a greater or lesser extent (Figure F.3). Various classifications have been published, but they tend to be based on layer thickness (stratified or laminated), the concentration and texture of debris (dispersed, solid, clotted) and the concentration and size of air bubbles (clear ice, bubbly ice).

FIGURE F.3 Basal ice facies in Fountain Glacier, Bylot Island, Canadian Arctic. Note the layering and variable debris content, as well as indications of deformation (July 2014).

Facies (glacier ice)

The types of ice making up the bulk of a glacier above the **basal ice** zone (Figure F.4). The principal ice type is coarse bubbly ice (white to pale blue), derived from metamorphism of snow to ice. Secondly, there is coarse clear (blue) ice, derived from water ice as in water-filled **crevasses** and extreme recrystallisation. Thirdly, the least common facies is fine-grained ice (white), which is the product of size-reduction of crystals under concentrated shear.

FIGURE F.4 The three main facies in glacier ice, exposed at the surface of Austre Brøggerbreen, northwestern Svalbard after cleaning. The dark ice is the coarse clear variety, the pale greyish ice at right is coarse bubbly, and the thin pale streaks are fine-grained ice (July 2013).

Facies (sediment)

The physical and chemical attributes and texture of the different types of sediment and rock. In glacial environments, there is a considerable array of sedimentary facies, collectively known as facies associations, which when rigorously described may be interpreted in terms of depositional process and overall environment respectively (Figure F.5).

Facies		Description	Interpretation
Clast-rich sandy diamicton		Non-stratified and poorly sorted, with up to 50% gravel clasts. mixed local and exotic clasts up to boulder-size, some of which are striated.	Derived from near-vertical foliation incorporating basal debris; melting out *in situ*.
Sandy gravel (muddy)		Non-stratified to weakly stratified, poorly sorted, with >50% clasts. Predominantly local clasts up to boulder size, some of which are striated.	
Sandy gravel		Non-stratified to weakly stratified, moderately well-sorted cobbles-granules in a sandy matrix. Mixed local and exotic clasts up to cobble-size.	Derived from foliated ice, but reworked by debris-flows and small streams within moraine.
Sand	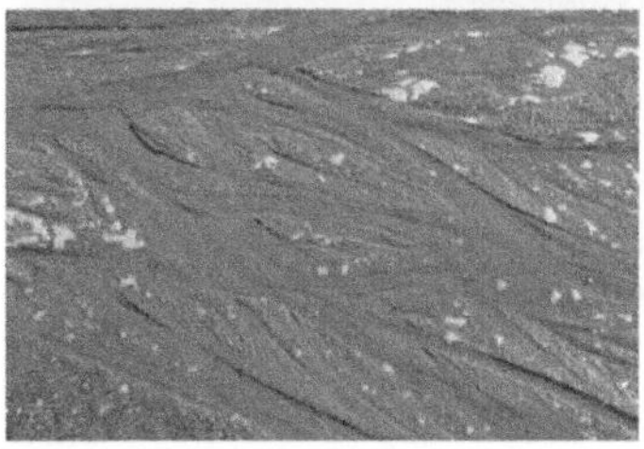	Patches of medium to coarse well-sorted sand on ice-shelf surface; found in stream-beds and ponds.	Aeolian debris accumulated on ice-shelf surface and concentrated by flowing water in late summer.

FIGURE F.5 Example of four sedimentary facies, with description and interpretation, in an ice-shelf moraine at the margin of George VI Ice Shelf, Antarctic Peninsula. (Modified from Hambrey, M. J., et al. 2015. *Journal of the Geological Society*, 134.)

Facies (snow)

Materials with distinct physical attributes resulting from the transition from snow to ice in the accumulation area of a glacier, and extending down to the **snow line** (Figure F.6). Dry-snow facies occur higher in the **accumulation zone** and experience little or no melting. Percolation facies represent the annual increment of new snow that is not completely wetted, the amount of percolation increasing with lower altitude. Soaked facies occur lower still, where the entire year's snowpack is raised to melting point and soaked with water. The lowest facies is the ablation facies where the snow cover is completely stripped away.

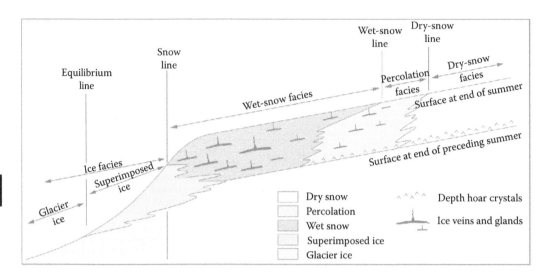

FIGURE F.6 The different snow facies at the transition zone from the accumulation area to the ablation area, showing the relationships with the snow and equilibrium lines. (Simplified from a diagram by C.S. Benson, US National Snow and Ice Data Center, http://nsidc.org/greenland-today/2013/02/greenland-melting-2012-in-review/)

Fast glacier flow (see glacier flow)

Fast ice

A type of **sea ice** that forms a continuous sheet attached to the shore, to a glacier front, or over shoals or between grounded icebergs (Figure F.7).

FIGURE F.7 Fast ice in a small bay north of Adelaide Island, western Antarctic Peninsula (October 2012). The inner bay with fast ice has trapped a number of icebergs. The grey ice to the seaward is newly formed spring ice, and beyond that are loose floes with new ice between.

Fault (glacier ice)

A displacement in a glacier formed by ice fracturing without its walls separating. It can be recognised by the discordance of layers in the ice on either side of the fracture (Figure F.8). **Crevasse traces** are a type of fault which do not show displacement. **Thrust-faults** are another variant.

F

FIGURE F.8   Faults with small-scale strike-slip displacements in Gulkana Glacier, Alaska Range, Alaska (June 2012).

Fault (sediment)

A displacement occurring in unconsolidated or frozen sediment. Such displacements are best observed in sand and gravel deposits, which are typical of **glaciofluvial sediments** (Figure F.9).

FIGURE F.9   Paired faults in cross-bedded sand and gravel, producing miniature graben and horst structures, Banc-y-Warren, Cardigan, southwest Wales. These sediments are the product of glaciofluvial deposition at the margin of the Irish Sea Glacier during the Late Glacial Maximum.

Felsenmeer (from the German for 'rock sea'; synonym, **Blockfield**)

Exposed bedrock surfaces that have been broken down by frost action to angular boulders (Figure F.10). These boulders may cover extensive areas and may be several metres deep. They are found in polar and high-mountain regions. Felsenmeer is commonly associated with areas that stand, or stood, above an icefield.

FIGURE F.10 Felsenmeer comprising Ordovician volcanic rocks at the summit of Scafell Pike, the highest mountain in England. *In situ* bedrock is visible on the left, and the blocks are up to several metres long.

Finger lakes

A collective term for a suite of narrow, roughly parallel lakes, carved out by glacial erosion beneath an **ice sheet**. The type locality is the Finger Lakes region of New York State, USA (Figure F.11).

FIGURE F.11 The Finger Lakes region of New York State, USA. An early December snowstorm emphasises the hills of central New York State. The hills contrast with the navy blue lakes. The original, roughly parallel river valleys and hills were shaped by ice sheets that attained depths of about 4 km during the last ice age. The basins containing the two largest lakes, Seneca and Cayuga, have been scoured to depths below sea level. The cities of Rochester, Syracuse and Ithaca are included in this field-of-view, taken from the International Space Station. (Astronaut photograph ISS010-E-9366, acquired 4 December 2004, provided by the ISS Crew Earth Observations experiment and the Image Science & Analysis Group, NASA.)

Firn (derived from the German; also known by the French term **névé**)

Dense, old snow in which the crystals are partly joined together, but in which the air pockets are still linked (Figure F.12). Firn usually refers to snow that has survived one melt-season or more. The density of firn ranges from 400 to 830 kg m^{-3}.

F

FIGURE F.12 Firn in the accumulation area of Triftgletscher, on Weissmies, Wallis, Switzerland. The top layer is snow from the previous winter, and the layers below are firn. Each year's firn accumulation amounts to several metres (Summer 1980).

Firn line

The line on a glacier that separates bare glacier ice from old snow (**firn**) at the end of the ablation season, or more specifically at the end of the **mass balance** year (Figure F.13). If a glacier is in steady state, the firn line coincides with the **equilibrium line**. If the glacier is experiencing a negative mass balance, the equilibrium line will be above the firn line, and *vice versa* for a positive mass balance.

FIGURE F.13 Aerial view of the firn line on Glärnischgletscher, Kanton Glarus, Switzerland. It almost coincides with the snowline, but the firn is evident from the dirty grey colour close to the lower edge of the snow (late summer 1982).

First-year ice

A type of sea ice that has no more than one year's growth. Thickness of a few centimetres to 2 m is typical of the polar seas. When undisturbed, the ice is smooth, but readily breaks up into floes, especially during the summer season (Figure F.14).

FIGURE F.14   Late spring first-year ice in Laubeuf Fjord, near Rothera Station (UK), Antarctic Peninsula. The mountains of Pourquoi Pas Island are in the background.

Fjord (from Norwegian; commonly spelt Fiord in North America and New Zealand)

A long, narrow arm of the sea, formed as a result of erosion by a **valley glacier** (Figure F.15a and b).

(a)

FIGURE F.15   The longest open fjords in the world are in Northeast Greenland. Viewed from the ground, they are spectacular, with walls up to 2000 m high above water level, but their dimensions are best appreciated in satellite images. (a) Kejser Franz Josef Fjord, Northeast Greenland, looking northwest towards the Greenland Ice Sheet, the source of the icebergs. The prominent mountain in the upper centre is Teufelschloss.   *(Continued)*

(b)

FIGURE F.15 (Continued)   (b) Overview of the complex fjord coastline of Northeast Greenland, spanning a distance from north to south of some 900 km. The Greenland Ice Sheet is on the left and cloud-covered ocean on the right, with mostly ice-free mountains and fjords in the intervening area. Kejser Franz Josef Fjord extends from the coast halfway along this stretch of coast. Scoresby Sund is the large embayment in the south. (Terra MODIS satellite image from NASA, taken on 21 August 2003.)

Floating tongue

A terminal part of a **glacier** that extends into the sea or a lake as a floating mass (Figure F.16). Lateral stress from valley sides, or from grounded zones, supports much of the weight of the floating ice, unlike in relatively unconstrained **ice shelves**.

FIGURE F.16   The floating tongue of Mackay Glacier, Victoria Land, Antarctica, with a skidoo party on sea ice in the foreground. This outlet glacier from the East Antarctic Ice Sheet flows past Mt England in the background (November 1997).

Flocculation

The process of contact and adhesion of particles suspended in a water body (Figure F.17). Particles are held together by a force analogous to the surface tension of a liquid, so they can be easily disaggregated. Silt and clay from glacial meltwater are prone to flocculate in saline water, creating larger particles that may behave physically as sand.

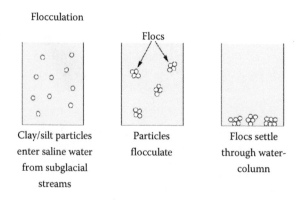

FIGURE F.17   The process of flocculation of clay and silt grains, and resulting settling.

Floe

A piece of loose floating sea, lake or river ice (Figure F.18). Some floes may exceed 10 km across, but most break up into metre-scale pieces.

FIGURE F.18 Thick floes of 'old' sea ice, beached at low tide near Terrapin Hill, Croft Bay, James Ross Island, Antarctica. The Royal Navy ship HMS Protector is in the background (March 2012).

Flow bands (see flowlines)

Flow law

The relationship between strain rate and shear stress in ice. The basic principles were established following experiments on blocks of ice in the laboratory, and the resulting relationship became known as 'Glen's Flow Law' after the British physicist who established it:

$$\dot{\varepsilon} = A\tau^n$$

where τ is the shear stress, $\dot{\varepsilon}$ the corresponding shear strain rate, n the creep exponent and A the flow parameter. The typical value for n in glaciers is approximately 3, but varies from 1.5 to 4.2. The parameter A depends on temperature and ice-crystal fabric. The application of this law to glaciers was subsequently generalised by Nye, another British physicist, following the first studies of deformation in a glacier borehole at Jungfraujoch, Switzerland.

Flowlines (also known as flow stripes, flow bands or streaklines)

These are longitudinal ice-surface features, particularly associated with **ice sheets**, and their constituent **ice streams** and **ice shelves** (Figure F.19). Flowlines are generally parallel to ice-flow direction, and individual flowlines may be traced for over 1000 km. They are visible both in areas of snow cover and bare glacier ice. There is evidence to suggest that flowline sets are the surface manifestation of the three-dimensional structure, **longitudinal foliation**, and that their relationship with flow direction may break down. Flowlines may also be evident in valley glaciers with strongly convergent flow; as snow cover melts during the ablation season, they are also revealed to be longitudinal foliation (Figure F.19c).

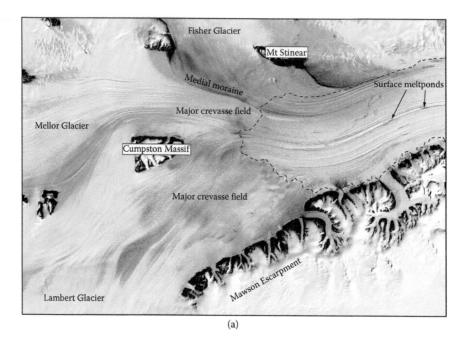

(a)

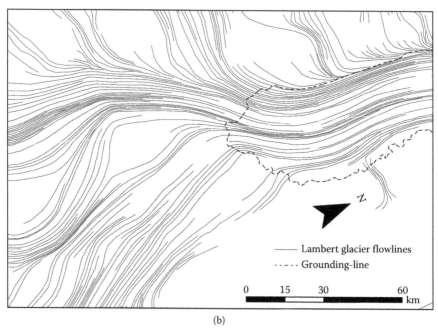

(b)

FIGURE F.19 Flowlines in an ice sheet and a valley glacier. (a,b). Flowlines at the confluence of three large ice streams in the East Antarctic Ice Sheet, the Lambert, Mellor and Fisher glaciers. (a) Is a Landsat image) and (b) is the structural interpretation. Note that the flowlines pass uninterrupted through major crevasse fields, across the grounding-line, and into an area of the Amery Ice Shelf dominated by surface ablation and visible surface melt ponds. The flowlines are interpreted as longitudinal foliation. (From Glasser, N.F., et al., *Earth Surface Dynamics*, 3, 239–249, 2015, published by the European Geosciences Union).

(Continued)

(c)

FIGURE F.19 (Continued) (c) Flowlines in the upper reaches of Tokositna Glacier, Denali National Park, Alaska. This down-glacier view shows topographically defined flowlines with ponds forming in the hollows. Flow convergence suggests that these flowlines are the surface expression of longitudinal foliation, buried under the snow (June 2012).

Flow stripes (see **flowlines**)

Flow unit

An individual component of glacier flow emanating from a distinct basin. In valley glaciers, multiple basins may result in several flow units merging, but their structural integrity can be recognised by medial moraines or strong **longitudinal foliation** at their boundaries, and by the preservation of arcuate structures within some of them (Figure F.20). Flow units are also recognisable in **ice sheets**, especially within their component **ice streams** and **ice shelves**, as defined by flowlines.

FIGURE F.20 Five flow units are visible in this aerial photograph of Fountain Glacier, Bylot Island, Canadian Arctic. The darker grey stripes, parallel to the edge of the glacier, are longitudinal foliation, and arcuate structures are visible with each flow unit (July 2014).

Fluctuations (see **glacier fluctuations**)

Flute

An elongate, smooth ridge of glacial sediment (till), typically tens of metres in length, a few metres wide and up to a couple of metres high (Figure F.21a). They occur singly or in groups, and collectively are known as **fluted moraine** (Figure F.21b). They are the created beneath an actively flowing glacier, and their orientation indicates the direction of flow. Ridges commonly form a tail behind an embedded boulder. The sediment is poorly sorted and mobile when wet, so they do not survive well after the glacier has receded. Flutes represent the smallest end member of the family under the heading **subglacial bedforms**.

(a)

(b)

FIGURE F.21   (a) A single flute, exposed in front of Trapridge Glacier, on a surface of basal till that was deposited in the 1950s. Note how the flute extends down-flow of an embedded boulder. (b) Fluted moraine, comprising basal till, in front of temperate glacier Austre Okstindbreen, northern Norway (August 1978).

Fluted snow

Snow on slopes with furrowed surfaces. (a) Compacted snow on lee slopes that is carved by strong winds (Figure F.22a). (b) Ribbed snow on steep mountain faces, formed by uneven ablation combined with small snow and ice avalanches (Figure F.22b).

(a)

(b)

FIGURE F.22 Fluted snow surfaces. (a) Metre-scale flutes on the south-facing slope of Roberts Massif at the edge of the East Antarctic Ice Sheet. They are carved by fierce winds from the Polar Plateau (January 1996). (b) A high-mountain ridge on Denali/Mt McKinley, Alaska, with fluted snow and ice extending down to the first crevasse (the bergschrund). Small avalanches are focused down the furrows (June 2012).

Flyggberg (derived from the Swedish)

Large asymmetrical hills of solid rock, resulting from glacial erosion. They are typically a few hundred metres high and over a kilometre long. A flyggberg is a large version of a **roche moutonnée**, and smaller forms may be superimposed upon them. The stoss side is glacially abraded and the lee side is glacially quarried (Figure F.23).

FIGURE F.23 Lembert Dome, a flyggberg (or large roche moutonnée) of bare granite, which rises 240 m above the Tuolumne River, Yosemite National Park, California, USA.

Fold

Layers of ice that have been deformed into curved forms by flow at depth in a glacier. The key attributes of a fold are its axial plane, fold axis, limbs and hinge (Figure F.24). Common types of fold in glaciers are **isoclinal**, **similar**, **chevron**, **sheath** (eyed) and **refolded**. They are illustrated separately. A wide variety of folds also occur in glacial sediments, formed especially when the sediment was wet and pressurised.

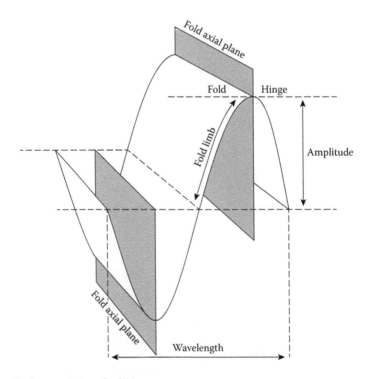

FIGURE F.24 The key attributes of a fold.

Foliation

Groups of closely spaced, often discontinuous, layers of coarse bubbly, coarse clear and fine-grained ice, formed as a result of shear or compression at depth within a glacier. Foliation occurs in most glaciers, and there are two main types: **longitudinal foliation** and **arcuate** (or transverse) **foliation** (Figure F.25). Foliation is commonly the product of transposition of earlier layering by increasingly tight folding, or as a new cleavage-like structure. Structural maps are produced to illustrate the three-dimensional arrangement of foliation and other structures in glaciers.

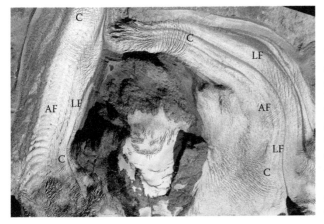

F

FIGURE F.25 Extract of composite aerial photograph of the middle reaches of Vadret da Morteratsch (left) and Vadret Pers (right), showing the two main types of foliation, longitudinal (LF), and arcuate (AF), which is associated with ogives, In addition, there are numerous crevasses (C). Ice flow is towards the top (north). The width of the image represents 3.3 km. The two glaciers were almost severed in 2015. (Photograph by Federal Office of Topography, Switzerland; Swisstopo.)

Forbes bands (see **ogives**)

Forefield

The zone in front of a glacier, usually pertaining to the area that has been recently **deglaciated** during recession. The forefield has a wide range of erosional and depositional features related to the glacier and its associated melt-streams (Figure F.26).

FIGURE F.26 Aerial image of the forefield of Austre Brøggerbreen, northwestern Spitsbergen, Svalbard. The height of the image represents 2.3 km, and the glacier is flowing north, towards the right. Key features are: the glacier snout on the left, an outer moraine complex at right, hummocky moraine, ice-cored moraine extending from medial moraines on the glacier, lakes with suspended sediment (orange), braided streams, and a subglacial meltwater channel at lower left. (Image from Natural Environment Research Council, UK, taken on 25 July 2004.)

Fountain

Water under pressure escaping at the surface of a glacier. The build-up of englacial and subglacial meltwater during periods of high melting or heavy rain can result in high water pressures in confined conduits in a glacier. Under such conditions, where englacial conduits intersect the surface at planes of weakness, such as fractures, a temporary fountain may occur (Figure F.27). See also **upwelling**.

FIGURE F.27   Fountain in the lower ablation area, close to the true left margin of Crusoe Glacier, Axel Heiberg Island, Canadian Arctic, on 10 July 2008. Water was emerging at a constant rate, the jet reaching a height of nearly a metre.

Fractures (ice)

Planar surfaces, often defined by planes of bubbles, resulting from stress release at the surface of a glacier surface. Most form normal to the maximum extending stress, but cross-fractures may also occur (Figure F.28). Some fractures may form in response to thermal changes.

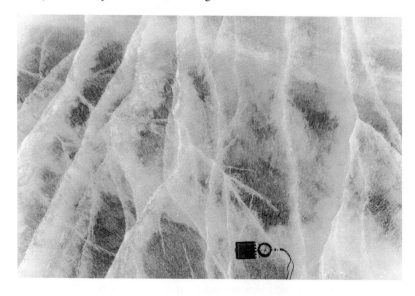

FIGURE F.28   Closely spaced fractures in coarse clear ice at the edge of George VI Ice Shelf. (November 2012).

Fractures (rock)

As ice overrides bedrock, the stress imparted may lead to fracturing, especially if the rock has previously been weakened by weathering (Figure F.29). Fractures tend to follow pre-existing structures in bedrock, such as bedding, foliation and joints. In certain circumstances, displacement of bedrock along thrust-faults may occur, producing **glaciotectonic structures**.

F

FIGURE   F.29   Heavily fractured bedrock in highly weathered gneiss in a former subglacial or ice-marginal meltwater channel, Fountain Glacier, Bylot Island, Canadian Arctic.

Frazil ice

Ice crystals that form as water freezes, especially if the water is supercooled by mixing at sub-zero temperatures. Frazil ice is common in marine waters beneath **ice shelves** in Antarctica, especially when mixed with fresher water from melting glacier ice. As frazil ice crystals grow, and develop an elongate form, they adhere to the base of an ice shelf, thereby contributing to its mass. If the surface of an ice shelf has a negative **mass balance**, frazil ice accreted at the base may emerge at the top surface, and the entire ice-shelf thickness may be composed of frazil ice (Figure F.30a and b). Frazil ice also forms beneath temperate glaciers by processes known as **glaciohydraulic supercooling** and **adfreezing**, where it forms an open network of platy crystals that can trap silt. Frazil ice also forms in open, turbulent supercooled water in the open ocean, rivers and lakes.

FIGURE F.30   Frazil ice in southern McMurdo Ice Shelf, Antarctica, formed by freeze-on to the base, and migration to the surface as a result of net surface ablation. (a) Layering of frazil ice that is typical of the surface ice structure in this part of the ice shelf.

(a)

(Continued)

F

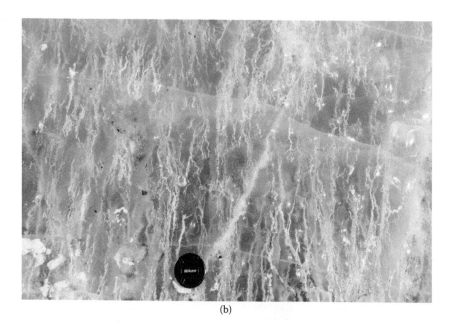

(b)

FIGURE F.30 (Continued)   (b) Elongate frazil ice crystals, vertically orientated, that make up the layers in (a) (November 2010).

Freeboard

The elevation of the surface of a floating body of ice (e.g. glacier tongue, ice shelf, iceberg), above the surface of the water (Figure F.31).

FIGURE F.31   An iceberg near Esperanza Station (Chile), east side of the northern Antarctic Peninsula. This iceberg has a freeboard of over 50 m, representing only 10–12% of its total thickness (February 2012).

Freshwater glacier

A **valley glacier** that terminates in a lake. Similar to a **tidewater glacier**, it generates **icebergs** and delivers large quantities of suspended sediment into the lake (Figure F.32).

FIGURE F.32 Aerial photograph of Nizina Glacier in the Wrangell Mountains of southern Alaska, which terminates in a large proglacial lake and produces numerous icebergs (July 2011).

Frictional heat

The heat generated in a **glacier** by **internal deformation**, sliding at the base, or deformation of **subglacial sediments**. This occurs when thermal energy is converted from gravitational potential energy as the glacier moves downslope. Heat generation is greatest when the shear **stress** at the bed of a glacier is large. If the ice is at the pressure melting point, internal shearing induces further melting. Heat generated by basal sliding and subglacial sediment deformation can be transmitted into the base of the glacier, aiding melting.

Frost action

The process of fracturing of rock by freeze–thaw mechanisms when water (e.g. from snow-melt) penetrates cracks in bedrock or boulders (Figure F.33). The process is particularly effective adjacent to glacier margins. Another process that produces similar weathering characteristics is **thermal stress fracturing**.

FIGURE F.33 A frost-shattered mudstone boulder adjacent to a glacier in Belemnite Valley, near Fossil Bluff Station (UK), Alexander Island, southwestern Antarctic Peninsula.

Frost flowers

Ice crystals that form on glacier surfaces, sea ice or lake ice in cold, calm conditions. Formation is favoured by air temperatures that are at least 15° cooler than that of the surface, or if conditions are foggy. The moisture in the air condenses over wide areas as **hoar frost**-like crystals on the surface (Figure F.34a). The ice crystals in frost flowers are usually dendritic (Figure F.34b), but they can also grow into rod shapes.

(a)

(b)

FIGURE F.34 Frost flowers on the surface of McMurdo Ice Shelf, December 2010, forming during calm, cold (−19°C) and foggy conditions. (a) A field of sparkling frost flowers before destruction by wind and sun. (b) Detail of individual frost flowers resting on dark ice. They are approximately 1–2 cm in length.

Furrowed lateral moraine

The near-vertical furrows and ridges that characterise **lateral moraines** in many alpine valleys. These moraines were formed at the valley sides during the **Little Ice Age**, and because the glaciers that created them have down-wasted, they may rise tens to a hundred metres above the ice surface (Figure F.35). The cohesive nature of the glacial deposits that make up these moraines allows steep (60–70°) cliffs to develop, but these are not stable and rainwater and snow-melt cause boulders and other material to be dislodged, forming the furrows.

FIGURE F.35   Furrowed lateral moraine on the left margin of Vadret da Morteratsch, with an alpine lodge, Boval Hütte, for scale (September 2012).

G

Gendarme

A rock pinnacle on a mountain, occupying and blocking a narrow ridge, specifically a glacially eroded **arête** (Figure G.1).

FIGURE G.1 A finger-like gendarme rises above the ridge that marks the Swiss–Italian border, to the right of the summit of Torrone Centrale (3290 m). The head of the Swiss glacier Vadrec del Forno is visible below (September 2011).

Geometrical ridge network

Low cross-cutting ridges in a glacier **forefield**. These rare features have been described from the area in front of a **surge-type glacier** as it recedes. The network comprises longitudinal ridges of basal glacial sediment that originate within debris-bearing **foliation** (foliation-parallel ridges), which are intersected by transverse ridges interpreted as **thrusts** (Figure G.2). Ridge height and width reach a few metres.

FIGURE G.2 Geometrical ridge network produced by a grounded part of Kongsvegen, which joins the tidewater glacier, Kronebreen, in the background. The ridge running towards the camera position is a foliation-parallel ridge, and is cut by the irregular thrust ridge in the background (July 1995).

Geothermal heat

The heat output from the Earth's interior. This affects glaciers especially in the Polar Regions, by warming the basal zone to the pressure melting point.

Gilbert-type delta

Named after G. K. Gilbert (1843–1918), an American geomorphologist, this type of delta is a wedge-shaped body of sand and gravel, consisting of relatively thick, prograding, steeply dipping foresets, overlain by thinner flat-lying topsets (Figure G.3). Flat-lying bottomsets occur are transitional with the base of the foresets. The sediment body is produced where a mountain river enters a lake, sometimes from a glacier source.

FIGURE G.3   Gilbert-type delta in cross-section, formed in the late Pleistocene Lake Bonneville before it shrank to the current Salt Lake in Utah.

Glacial breach (see **breached watershed**)

Glacial buzzsaw effect

The effect of glaciers eroding mountains that are undergoing tectonic or **isostatic uplift**. As mountains increase in height, they allow more intense glacial erosion, especially above the glacier's **equilibrium line**. This limits the amount by which the mountains can grow. This 'buzzsaw effect' can only operate if the glaciers are wet-based; in cold regions, where the ice is frozen to the bed, glaciers form a protective cover. Note: a 'buzzsaw' (North American English) is a power saw with a rapidly rotating toothed disc, known as a circular saw in British English.

Glacial erosion

Erosion of the landscape by the processes of bedrock **abrasion** and rock fracture (**quarrying**), or the removal of soft sediment, at the base of a glacier or ice sheet (Figure G.4). The effects of meltwater commonly supplement these processes. Erosional forms occur at a variety of scales, ranging from **polish** and **striations** at the small scale, through **channels** and **grooves** at the intermediate scale, to **troughs** and **cirques** at the large scale.

FIGURE G.4 Glacial erosion of bedrock near the margin of Kongsbreen (west) on Ossian Sarsfjellet, northwestern Spitsbergen, Svalbard. The sunlit surface is abraded and carries striations. The shadowed surfaces were fractured prior to debris removal (i.e. quarried).

Glacial lake

Water bodies collect in a variety of subglacial, supraglacial, englacial, ice-marginal environments, where an ice, sediment or rock barrier impedes drainage. Glacial lakes vary considerably in size: from ponds to the vast expanses of water that prevailed at the margins of the last great Northern Hemisphere ice sheets. Glacial lakes occupying rock basins tend to be stable, those dammed by ice are prone to regular drainage (commonly annually), whereas those dammed by moraine undergo a single catastrophic failure. Lakes are classified according to their topographic relationship with a glacier, and the nature of the dam material. They include the following, which are illustrated separately: **proglacial lake**, **moraine-dammed lake**, **ice-dammed lake** and **subglacial lake**.

Glacial lake outburst flood (commonly abbreviated to GLOF)

During a period when a glacier recedes from its terminal moraine, a lake may form, impounded by an unstable pile of debris and buried ice (Figure G.5). Catastrophic failure of the moraine will result in a potentially devastating flood. These events are usually associated with high-mountain regions such as the Andes and Himalaya (*cf. Jökulhlaup*, which is the failure of an ice-dammed lake).

FIGURE G.5 The aftermath of a moraine-dammed lake failure at Chukhung Glacier, Khumbu Himal, Nepal. The lake had a maximum volume of 5.5×105 m^3 and drained as a result of breaching of the terminal moraine, visible in the middle of this photograph. An estimated 1.3×105 m^3 of material was removed from the terminal moraine during breach development, and this is spread out in the sediment fan below (May 2003).

Glacial lake shorelines

Shorelines left at the margin of an **ice-dammed lake** owing to fluctuations in water level. Partial filling of, or periodic drainage from, these lakes may give rise to multiple shorelines (Figure G.6a). If successive shorelines are relatively stable for several years, wave-cut notches in bedrock, as well as beach gravel accumulations may occur (Figure G.6b).

(a)

(b)

FIGURE G.6 Example of modern and ancient glacial lake shorelines. (a) Drained ice-dammed lake and multiple shorelines, near the confluence of Blårevbreen (foreground) and Oslobreen (background) in northeastern Spitsbergen, Svalbard (August 2010). (b) A series of three prominent shorelines, known as the 'Parallel roads', resulting from successive failures of ice dams in Glen Roy, Grampian Highlands, Scotland. These shorelines date from the Younger Dryas period, about 11,000 years ago. An ice-dam blocked the valley at the position of the photo-grapher, and water escaped over different cols when successively unblocked by ice. These shorelines form notches in bedrock, and so were relatively long-lived.

Glacial landsystem (see landsystem)

Glacial period/glaciation

A period of tens of thousands of years when large areas of the Earth (including present-day temperate latitudes) were covered by glacier ice. Numerous glacial periods have occurred within the last few million years and are separated by **interglacial periods**. Glacial periods have also occurred sporadically throughout geological time.

Glacial sedimentary logging

The procedure for recording the different sediment types (**facies**) in the form of vertical columns. The thickness of the layers is recorded, along with pertinent sedimentary structures, and facies are interpreted in terms of process of deposition (Figure G.7).

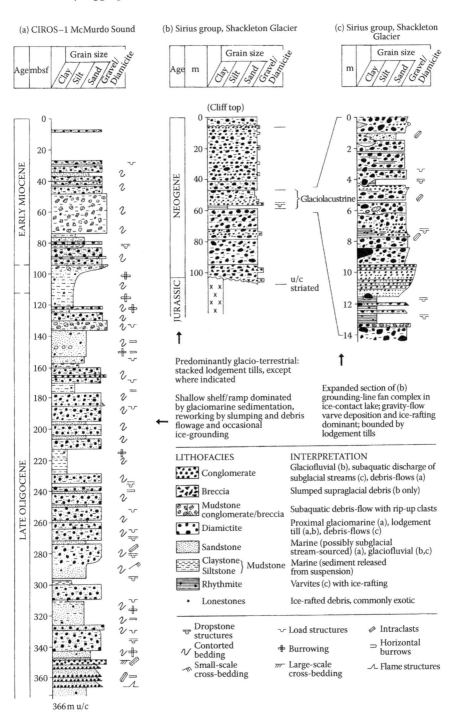

FIGURE G.7 Examples of sedimentary logs through thick glacial sequences from Antarctica at different scales. Log (a) is the upper part of a 702-m-long core drilled into the sea bed in McMurdo Sound. Log (b) is at the same scale from one of the southernmost terrestrial outcrops in the world, at Shackleton Glacier. Log (c) is an expanded version of several metres of section from (b) to illustrate the complexity of the record. Each of the different sediment types (facies) is interpreted in terms of depositional setting. (Reproduced from Hambrey, M.J. and Glasser, N.F., Glacial sediments: processes, environments and facies, In Middleton, G.V. (ed.) *Encyclopedia of Sediments and Sedimentary Rocks*, Kluwer, Dordrecht, the Netherlands, 316–331, 2003.)

Glacial stream

A stream emanating from a glacier and continuing down valley, commonly spreading out on a **braid plain** (Figure G.8).

FIGURE G.8   A glacial stream emerging from an ice cave at the snout of Franz Josef Glacier, South Island, New Zealand (September 2008).

Glacial surface of erosion (see **sequence stratigraphy**)

Glacial trough

A glaciated valley or fjord, commonly characterised by steep sides and a flattish bottom, with multiple basins, resulting primarily from abrasion by strongly channelled glacier ice, for example, a **valley glacier** or **outlet glacier**. A glacial trough has a **parabolic form** in cross-section (Figure G.9), sometimes with sides being steep enough to form a **U-shape**.

FIGURE G.9   The classic glacial trough of Glencoe in the Grampian Highlands of Scotland, with its parabolic cross-section. A glacially abraded pavement with an erratic is in the foreground.

Glacially transported boulder

A boulder that has been transported at the bed or on the surface of a glacier, and deposited some distance from its source. More specifically, if the boulder is of a different rock type from the local bedrock, it is referred to as an **erratic** (Figure G.10).

FIGURE G.10 A glacially transported boulder of gneiss on the floor of the valley of the glacier, Vadret da Morteratsch, Switzerland. This is the same rock type as the surrounding bedrock. The partial rounding and striations on the flatter surface indicate transport at the base of the glacier.

Glaciated

The character of land that was once covered and modified by glacier ice in the past, but no longer at the present time (*cf.* **glacierised**) (Figure G.11).

FIGURE G.11 The Yosemite Valley (with Half Dome on the right) and the adjacent highlands in California represent a glaciated highland landscape. Ice covered these highlands and the valley was the focus of a vigorously eroding outlet glacier.

Glaciated landscape

A landscape that has been influenced by **glacial erosion** and **glacial deposition**, producing a wide range of landforms. Glaciated highland landscapes are dominated by erosional forms such as **pyramidal peaks**, **arêtes**, **cirques**, **glacial troughs** and **fjords**, along with volumetrically small deposits of glacial and **glaciofluvial sediment** (Figure G.12a and b). Glaciated lowland landscapes are usually dominated by depositional landforms, such as sheets and mounds of glacial deposits in a wide range of geometric forms.

(a)

(b)

FIGURE G.12   A glaciated highland landscape, the Lake District, northwest England. (a) The overdeepened lake basin containing England's deepest lake, Wastwater. The lake is surrounded by the Ordovician volcanic Central Fells including Great Gable, a pyramidal peak left of centre, and Scafell to the right. The foreground is occupied by Ordovician granite roches moutonnées. (b) The view from near the summit of Scafell Pike, England's highest peak, looking past a cirque basin and glacial troughs to Great Gable, right of centre.

Glacier

A perennial mass of ice, irrespective of size, originating on land by the recrystallisation of snow to firn and then ice. A glacier also shows evidence of past or present flow. Glaciers are classified by size and morphology including **ice sheet** (at the largest scale), **ice cap** or **ice dome**, **glacier complex**, **highland icefield**, **alpine** (or **valley**) **glacier**, **cirque glacier**, **ice apron** or **mountain apron glacier**, **niche glacier**, **hanging glacier**, **reconstituted/regenerated/rejuvenated glacier**, **expanded foot/piedmont glacier** and **glacieret**. Glaciers which interact with water bodies have generated several additional names, including **ice shelf**, **calving glacier**, **tidewater glacier** and **freshwater glacier**. Glaciers may also be classified according the temperature distribution within them, that is, their thermal regime. A glacier that is at the pressure melting point throughout is referred to as a **temperate** or **warm glacier**. A **cold** or **dry-based glacier** has ice that is below the pressure melting point, and is frozen to its bed. A **polythermal glacier** is a hybrid of the two types, typically having warm ice at the base where it is thickest (as a result of warming from geothermal heat), whilst the margins and upper parts are cold. The terms **polar glacier** and **subpolar glacier** were previously used to describe cold and polythermal glaciers, respectively, but these terms are no longer favoured because of their geographical connotations. All these glacier types are described and illustrated separately.

Glacier (or glacial) advance

Increase in the length of a glacier, as measured from a fixed point. If the glacier terminates on land, the fixed point is in front of the glacier, but for **tidewater glaciers**, advances (and recessions) are measured from the sides. Larger ice masses require the use of repeat satellite images to determine such fluctuations. Advances are typically associated with a cooling climate, gains of mass through increased snowfall in the accumulation area, a lengthening of the accumulation season or a shortening of the melt-season (Figure G.13a and b).

(a)

FIGURE G.13 Recent glacier advances on decadal and yearly time scales. These are repeat photographs from the same positions. The vertical cliffs at the snout of each glacier are characteristic of advancing glaciers. (a) Crusoe Glacier, a polythermal glacier on Axel Heiberg Island in the Canadian Arctic has been advancing slowly since records began (in the 1960s). The glacier has advanced a few dozen metres, and also thickened (upper: July 1977, lower: July 2008).   *(Continued)*

(b)

FIGURE G.13 (Continued) (b) A recent advance of the snout of a temperate valley glacier, demonstrated by Fox Glacier, New Zealand (left: April 2006; right: April 2007). Steeply graded, highly dynamic glaciers like this respond rapidly to mass balance changes. This glacier is now (in 2015) receding rapidly.

Glacier bed

The surface beneath a glacier, where most of the erosional and depositional processes take place. Glacier beds may be hard (bedrock) or soft (sediment), and wet or dry, depending on the temperature of the ice and availability of meltwater.

Glacier–climate interaction

Glaciers are sensitive indicators of climate change, in that they respond to changes in temperature and precipitation on decadal and centennial time scales. Contemporary changes are manifested in **advance** or **recession** of the glacier **terminus**, and tend to even out year-to-year variations. Investigating the response of glaciers to climate change requires a combination of **mass-balance** measurements and an assessment of meteorological variables (Figure G.14). On the large scale, **ice sheets** are not only influenced by climate change, but they

FIGURE G.14 A meteorological station on Ghiacciaio del Forni in the Italian Alps, established to measure the response of the glacier to climatic change (July 2007).

themselves also influence climate by their reflectivity of radiation. Ice sheets also record past climate, which can be obtained by ice-coring. These records span up to a million years in Antarctica.

Glacier complex

A number of interconnected glaciers, linked across the ice equivalent of watersheds (**ice divides**) (Figure G.15).

FIGURE G.15 A glacier complex, with a highland icefield feeding numerous valley glaciers in northern Ellesmere Island (July 2011).

Glacier dynamics

Refers to how the motion of glaciers and ice sheets varies in space and time (Figure G.16). Many variables are involved, including **mass balance**, **thermal regime**, **glacier hydrology**, basal conditions, structure and interaction with the sea or lakes.

FIGURE G.16 Highly dynamic (fast-flowing) glaciers entering Kongsfjorden, northwestern Spitsbergen, Svalbard. The photograph is taken from the calving front of Conwaybreen, looking towards another calving tidewater glacier, Kongsbreen (west arm) (July 2009).

Glacier-fed delta

A body of sediment, supplied by a glacier stream but separated from the glacier, that is deposited in a lake (Figure G.17). The top surface of the delta is usually braided, with frequently migrating channels, and the sediment-laden stream generates plumes that extend far into the lake. **Sand** and **gravel**, from which the delta is constructed, are the dominant sedimentary **facies**.

FIGURE G.17 Delta at the head of Lake Louise, Banff National Park, Alberta, Canada. Victoria Glacier, at the head of the valley, feeds this lake. Suspended sediment from the glacial stream gives rise to the turquoise colour in the lake.

Glacier flea

The glacier flea, *Desoria saltans*, is a hexapod from the class of Collembolans or springtails (Figure G.18). It is 1.5 to 2.5 mm long, and lives on alpine glaciers, and may be particularly abundant beneath boulders resting directly on glacier ice. Its main food source is **cryoconite.** When disturbed, glacier fleas can escape by means of a dorsal appendix, which lets them jump a considerable distance. This has led to their common name, although they are not fleas. Various sugars protect them from freezing, even at temperatures of between −10 and −15°C. Their preferred ambient temperature is 0°C.

FIGURE G.18 The glacier flea, *Desoria saltans*, observed after removal of a boulder on the surface of Vadret da Morteratsch, Switzerland (October 2012).

Glacier flow/motion

Glaciers flow by three main modes: internal deformation, basal sliding and movement on a deformable bed of sediment (Figure G.19a). Fast glacier flow (hundreds of metres per year) is characterised by a heavily crevassed surface. Slow glacier flow (a few metres per year), by a well-developed surface stream network and a lack of crevasses (Figure G.19b). See also **glacier dynamics.**

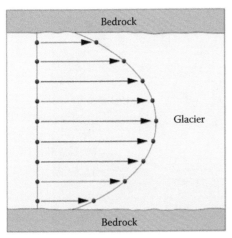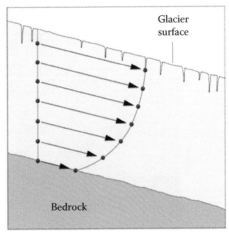

Plan view   Longitudinal cross-section

(a)

(b)

FIGURE G.19 (a) Relationship between two of the three main components of glacier flow in plan view and in a longitudinal cross-section. This relationship is typical of a temperate valley glacier. (b) Contrasting glacier flow types in a converging glacier system in northwestern Spitsbergen, Svalbard. Fast flow is indicated by the heavily crevassed Kronebreen; this glacier is thought to be the fastest steadily flowing glacier in Svalbard, with a maximum velocity of 700 m per year. Slow flow is indicated in Kongsvegen at lower right, which has a history of surging, but is currently in a quiescent state, moving only a few metres per year (Summer 1999).

Glacier fluctuations

The changes that occur in glaciers in terms of length (position of the terminus), area and mass balance (Figure G.20). These changes are documented for many glaciers globally by the World Glacier Monitoring Service, based at the University of Zürich.

FIGURE G.20 The fluctuating snout of Steigletscher, Switzerland, showing both an advance and retreat. The photos are from: 1982, 1985, 1991, 1994, 1999 and 2006 in the order left column top to bottom, then right column.

Glacier forefield (see **forefield**)

Glacier front (see **glacier terminus, snout**)

Glacier health

Refers informally to whether a **glacier** has a positive or negative **mass balance**. Most of the world's glaciers in 2015 are 'unhealthy', characterised by gently graded, crevasse-free snouts that are receding (Figure G.21a). 'Healthy' glaciers generally have a steep, heavily crevassed snout that is advancing (Figure G.21b).

FIGURE G.21 Contrasting 'unhealthy' and 'healthy' glaciers. (a) Unnamed receding glaciers near Trapridge Glacier, Yukon, Canada (July 2006). *(Continued)*

(a)

(b)

FIGURE G.21 (Continued)   (b) Advancing Fox Glacier, South Island, New Zealand, with push moraine in front, April 2006; this glacier has since gone into rapid recession.

Glacier hydrology

The study of water flow on, within and at the base of a glacier, as well as the glacier's discharge characteristics (Figure G.22).

FIGURE G.22   Aerial view of the surface of Gates Glacier, Wrangell Mountains, Alaska, demonstrating some aspects of the complex hydrology of a valley glacier: supraglacial streams and lakes, and stream terminations at moulins (July 2011).

Glacier ice

Any ice within, or originating from, a glacier, whether on land or floating on the sea as **icebergs**. Glacier ice is the product of recrystallisation or metamorphism of snow, *via* **firn** to ice, or the refreezing of meltwater, to a point where few voids remain connected (Figure G.23a). The density at which firn becomes glacier ice is approximately 830 kg m^{-3}. Subsequent weathering may lower the density of the ice as a result of preferential melting along crystal boundaries (Figure G.23b).

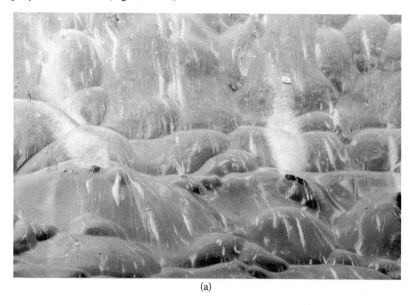

(a)

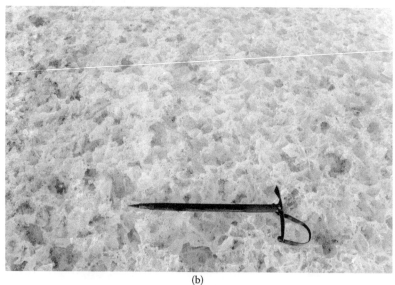

(b)

FIGURE G.23　Fresh and weathered glacier ice from southern Iceland. (a) Wave-washed glacier ice in small iceberg, showing pale blue bubbly appearance, beach near Jökulsárlón. (b) Weathered glacier ice at snout of Breiðamerkurjökull, showing large ice crystals (both images taken in March 2015).

Glacier karst

Debris-covered **stagnant ice**, sometimes found at the snout of a receding glacier, with numerous depressions, lake-bearing caverns and tunnels (Figure G.24).

FIGURE G.24 Glacier karst in the lowest part of Kennicott Glacier, Wrangell Mountains, Alaska (July 2011).

G

Glacier lassitude

The feeling of extreme tiredness induced by a combination of hot sun and still air on a glacier (Figure G.25). This combination results in saturation of the air, so that heat loss from the body is disrupted. It is most commonly experienced on high-altitude glaciers in temperate regions.

FIGURE G.25 Experiencing glacier lassitude on Stagnation Glacier, Bylot Island, Canadian Arctic (July 2014).

Glacier-like forms (Mars)

Landforms on Mars that are predominantly composed of ice–dust mixtures and are visually similar to terrestrial **valley glaciers**. They show signs of downhill viscous deformation and an expanded former extent (Figure G.26). They have distinct linear features reminiscent of **flowlines** on glaciers on the Earth. They also show ridges, interpreted as **moraines**, and a wide range of other geomorphological features.

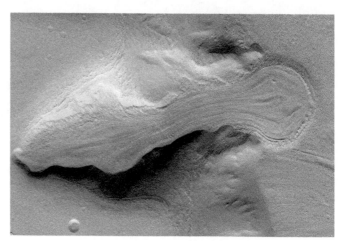

FIGURE G.26   A martian glacier-like form with distinctive structural attributes: view of the Protonilus Mensae–Coloe Fossae region of the northern mid-latitudes on Mars, from a Thermal Emission Imaging System (THEMIS) image mosaic (Dickson, J.L., Head, J.W., Marchant, D.R. 2008. Late Amazonian glaciation at the dichotomy boundary on Mars: evidence for glacial thickness. *Geology*, 36, 411–414.). Ice flow is from left (north) to right from an inferred ice cap. The lobe is approximately 2 km wide.

Glacier margin

The line separating a glacier (including connected debris-covered ice) from ice-free terrain (Figure G.27).

FIGURE G.27   Margins of Muldrow Glacier, Denali National Park, Alaska, June 2012. The glacier margin, mostly covered by dark grey rockfall debris, is clearly defined against the steep valley sides. Glacier flow is towards the top right.

Glacier mice (see *Jökla-mýs*)

Glacier milk

Meltwater from glaciers, which commonly has a milky appearance as a result of suspended fine sediment (Figure G.28).

FIGURE G.28   Outflow from Fox Glacier, South Island, New Zealand, April 2008. The pale blue 'milky' appearance is from suspended sediment derived from metamorphic rocks.

Glacier portal

The large cave that sometimes develops at the glacier **snout**, at the site of the main glacial stream discharge (Figure G.29). These caves are unstable, as their roofs are prone to collapse.

FIGURE G.29   Glacier portal at Vadret da Morteratsch, Grisons, Switzerland, in August 2011, before this entire lower part of the glacier disappeared.

Glacier recession (retreat)

The loss of mass from a glacier in response to a negative **mass balance**. Recession involves reduction in length of the glacier, but is also reflected in lowering of the glacier surface (Figure G.30).

(a)

(b)

FIGURE G.30 Repeat photographs to show the substantial recession of Kongsbreen, a tidewater glacier in northwestern Spitsbergen, Svalbard, over a 13-year period. Images date from July 1996 and July 2009. Note the encroachment of the marine embayment, the exposure of the lake, removal of the medial moraine and substantial thinning.

Glacier runoff

Refers to all the sources of water that flow off and from a glacier (Figure G.31). The greatest volume of water comes from subglacial or lateral melt-streams, but this is supplemented by runoff of small streams directly from the glacier surface.

FIGURE G.31 A warm day on Bylot Island, Canadian Arctic, in July 2014, resulting in substantial runoff from the surface of Fountain Glacier *via* supraglacial streams.

Glacier sole

The lower few metres of a (usually sliding) **glacier** that is either debris-poor, such as beneath a **temperate glacier** (Figure G.32), or debris-rich as at the base of **polythermal** or **cold glacier**.

FIGURE G.32 The relatively clean sole of lower Grosser Aletschgletsher, Switzerland. This is viewed from within a cave at the glacier margin (October 2010).

Glacier table

A boulder perched on a pedestal of ice, formed when ablation of ice beneath the boulder is less than ablation of the surrounding ice (Figure G.33). This process occurs during periods of strong solar radiation. Many boulders tilt towards the maximum input of radiation, and eventually slide off their pedestals.

FIGURE G.33   The authors standing on a granite glacier table on Vadrec del Forno, Switzerland (September 2011).

Glacier terminus (also referred to as glacier front, snout)

The lowermost end of a **glacier**, whether ending on land or in a water body (Figure G.34).

FIGURE G.34   The irregular terminus of Rob Roy Glacier, South Island, New Zealand. The glacier terminates on a steep rocky slope and has several individual snouts (January 2011).

Glacier tongue (or ice tongue)

1. An unconstrained, floating extension of an **ice stream** or **valley glacier**, projecting into the sea (Figure G.35a). The margins commonly have a series of 'fins', which stand tens of metres above the waterline.
2. The lower, laterally confined part of a **valley glacier** or outlet glacier, usually longer than wide, and broadly coinciding with the **ablation area** (Figure G.35b).

(a)

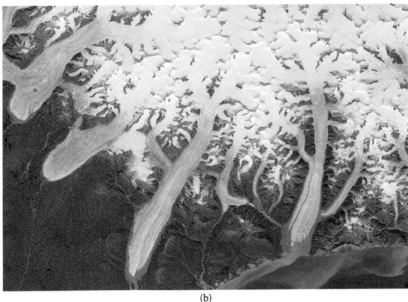

(b)

FIGURE G.35 (a) Erebus Glacier Tongue in McMurdo Sound, Antarctica, which flows off the volcano of the same name and floats on the sea. The tongue in this January 1996 image is here surrounded by sea ice. (b) Satellite image of several glacier tongues on Bylot Island, Canadian Arctic, extending from a central highland icefield. Fountain Glacier, featured in several entries in this book, is the small hook-shaped glacier in the lower middle part of the image (August 2013 NASA Landsat image no. LC80330082013221LGN00_B1_B2_B3_TIF).

Glacier vegetation

Colonisation of glaciers by vegetation takes places readily on stagnant debris-covered areas, especially if there is a plentiful source of fine sediment (Figure G.36). Some glaciers are able to support shrubs, alpine flowers and grasses (as in the Alps), while others support well-developed forests (as in Alaska). Slow melting of the underlying dead ice may cause the vegetation to destabilise.

FIGURE G.36 Early colonisation of stagnant ice by willow, at the snout of Unteraargletscher, Switzerland (Summer 2004).

Glacier–volcano interaction

The close association of volcanoes and glaciers, usually found at constructive and destructive plate margins (Figure G.37). Where glaciers grow on volcanic edifices, and eruptions take place, a suite of unique sediments and landforms are produced, including **jökulhlaup** deposits, **debris flows** and subglacially produced lava and breccia (or **hyaloclastite**).

FIGURE G.37 Glacierised summit plateau of the volcano Nevado del Ruiz in Colombia, located on the destructive plate margin that uplifted the Andes. This photograph was taken in November 1985, after the catastrophic eruption on 13th October. Lahars, caused by the eruption, generated debris flows that killed nearly 20,000 people in the city of Armero. (Photograph by US Geological Survey.)

Glacieret

A very small **glacier**, typically less than 0.25 km² in extent, with limited indications of flow. A glacieret is defined as being more than two consecutive years old. Glacierets may be of any shape, and usually occupy sheltered hollows in the landscape (Figure G.38a). Wind-blown snow may be the main contributor to such glaciers. A glacieret may also be the remnant of a much larger glacier (Figure G.38b).

(a)

(b)

FIGURE G.38 (a) Glacieret in southern Axel Heiberg Island, Canadian Arctic. Note that the high plateau behind is not glacierised. The glacieret was formed as a result of wind-blown snow accumulation in the hollow facing the camera. However, little snow remains on the glacier which has a severely negative mass balance (July 2008). (b) Gärstengletscher, a glacieret on the western side of the Gärstenhörner (3184 m) near Grimsel Pass, Switzerland. This glacier is a remnant of a larger cirque glacier (August 2012).

Glacierised/glacierised

The state of the terrain covered by glacier ice at the present day (Figure G.39) (*cf.* **glaciated**).

FIGURE G.39   Heavily glacierised terrain: part of the Antarctic Peninsula Ice Sheet in Palmer Land, where outlet glaciers are flowing into George VI Ice Shelf (November 2012).

Glacigenic

Of glacial origin, generally referring to sediment or landforms with a greater or lesser imprint of glacier transport or erosion.

Glacio-eustacy

Global changes in relative sea level arising from the growth and decay of terrestrial glaciers and ice sheets.

Glacigenic/glaciogenic sediment flow (see **debris flow**)

Glaciofluvial/glacifluvial sediment

Sediment that is carried and reworked by steams emanating from a glacier. The sediment is moderately well sorted and consists predominantly of **sand** and **gravel**, commonly cross-bedded and formed into **bars** (Figure G.40).

(a)

(b)

FIGURE G.40 (a) Glaciofluvial sediment comprises this sand and gravel bar in a glacial stream, which emanates from Hoffellsjökull, northwest of Höfn, Iceland (March 2015). (b) Glaciofluvial gravels just below Fox Glacier, Southern Alps, New Zealand. A large amount of suspended sediment is being carried by the river, between the mobile bars that comprise gravel (January 2011).

Glaciogenic (see **glacigenic**)

Glaciohydraulic supercooling (see **adfreezing**)

Glacioisostacy (ice-sheet loading)

The effect of an ice sheet loading the Earth's crust, resulting in displacement of the underlying mantle. This affects relative sea level adjacent to areas which support an ice mass. However, global sea level is not affected. See also **isostacy**.

Glaciolacustrine/Glacilacustrine delta

A delta formed by the discharge of a glacial stream into a nearby lake. It comprises topsets, cross-sets and bottomsets of sand and gravel (Figure G.41) (*cf.* **Gilbert-type delta**).

FIGURE G.41   Partly eroded glaciolacustrine delta, with topsets and cross-sets visible, Adams Inlet, Glacier Bay, Alaska.

Glaciolacustrine/Glacilacustrine sediment

Sediment that is deposited in a glacially influenced lake. It typically comprises couplets of **sand** and **mud**, or silt and clay, representing summer and winter settling out of sediment from suspension, respectively. If known to be annual, couplets are referred to as **varves**. Subaqueous **debris flows**, **dropstones** and current-winnowing may interrupt the laminae (Figure G.42a and b).

(a)

(b)

FIGURE G.42 Different styles of glaciolacustrine sedimentation. (a) Strongly laminated glaciolacustrine sediment of the pre-Quaternary age at Bennett Platform, Shackleton Glacier, Antarctica. These several million year old sediments are indicative of a warmer and wetter climate than that of today. (b) Less well-developed glaciolacustrine sediment belonging to the Last Glacial Maximum (Devensian), and disrupted by debris flows and iceberg rafting, Tonfanau, Gwynedd, Wales. Pocket knife for scale.

G

Glaciomarine/glacimarine sediment (also **glacial marine sediment**)

Sediment that originates from a calving glacier that terminates in the sea (Figure G.43a and b). Several sedimentary types (**facies**) that have glacial characteristics occur in the marine environment, notably (a) massive and stratified diamicton, interpreted as the product of rain-out of glacial sediment close to the grounding-line, or as **debris flows**; (b) couplets of **sand** and **mud** (known as **cyclopsams**), or silt and clay (**cyclopels**) punctured by outsized stones, representing tidally influenced changes in sedimentation, and containing ice-rafted dropstones.

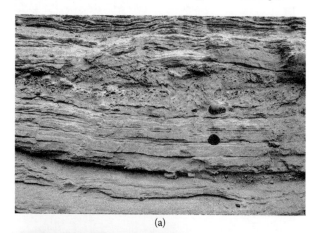

(a)

FIGURE G.43 Contrasting glaciomarine sedimentary facies. (a) Glaciomarine sediment on Whidbey Island, Washington State, USA, dating from the Last Glacial Maximum (the Wisconsinan Glaciation). A debris flow of diamicton, formed between rhythmites, or specifically, tidally influenced cyclopsams. Lens cap (5 cm) for scale. (b) A thick unit of slightly stratified diamictite within the Pagodroma Group, Fisher Massif, East Antarctica. This exposure forms part of a very thick sequence of sediment deposited close to a glacier in a fjord during Miocene time.

(b)

Glacio-speleology

The sport of ice caving. Glaciers have similar types of cave systems eroded by meltwater to those in limestone areas, including vertical shafts (**moulins**), **englacial channels**, and **subglacial tunnels** and **caves**. A number of glaciologists undertake glacio-speleological research, and they have provided considerable information about water circulation, both within glaciers (Figure G.44a) and the Greenland Ice Sheet (Figure G.44b).

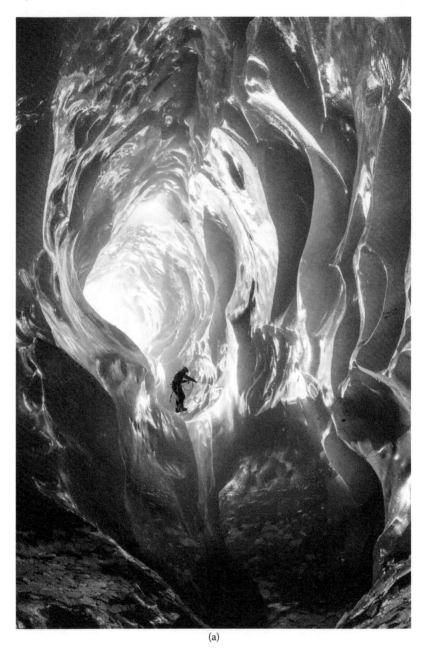

(a)

FIGURE G.44 Glacio-speleological research in action. (a) A glaciologist undertaking a roped traverse in a subglacial tunnel beneath Gornergletscher, Switzerland, in November 2012. Bedrock is visible in the foreground. The entrance to this tunnel was at the glacier margin, some 50 m behind the camera position.

(Continued)

(b)

FIGURE G.44 (Continued)   (b) Exploring a 10 m diameter moulin formed by hydraulic fracture through the Greenland Ice Sheet during the rapid drainage of a supraglacial lake (August 2010). (Photographs by Sam Doyle.)

Glaciotectonism/glacitectonism (or **glaciotectonic deformation**)

The process whereby subglacial and proglacial sediment or bedrock is disrupted by ice flow. Sediment commonly reveals landforms with discrete topographic features, within which folds and thrusts are commonplace (Figure G.45a). Bedrock is subject to brittle fracture with faults and thrusts being typical (Figure G.45b).

(a)

FIGURE G.45   Effect of glaciotectonism on sediment and bedrock. (a) Strongly deformed lacustrine and fluvial sediment in a terminal moraine enclosing Lake Pukaki, South Island, New Zealand. This structure was formed by an expanded Tasman Glacier at the Last Glacial Maximum.   *(Continued)*

(b)

FIGURE G.45 (Continued)   (b) Glaciotectonically disrupted weathered gneiss near the snout of Fountain Glacier, Bylot Island, Canadian Arctic. Several thrusts and other fractures rise up from a sole thrust, which is prominent dark grey zone of fractured bedrock in the lower third of the photograph.

Glaze (see **verglas**)

Glen's flow law (see **flow law**)

GLOF (see **glacial lake outburst flood**)

Gravel

A general term for all sedimentary material with particle size larger than 4 mm (Figure G.46). Gravel embraces pebbles (4–64 mm), cobbles (64–256 mm) and boulders (>256 mm).

FIGURE G.46   Recently deposited supraglacial gravel, dominated by cobbles and boulders near the snout of Vadret da Morteratsch, Switzerland (September 2011).

Groove

A glacial abrasional form, with **striated** sides and base, orientated parallel to the ice-flow direction, and commonly several metres wide and deep (Figure G.47).

FIGURE G.47   Groove, measuring 3 m deep and 8 m wide, with striated sides, Whitefish Falls, near Sudbury, Ontario, Canada. This groove was formed under the Laurentide Ice Sheet (Wisconsinan Glaciation).

Grounding-line/zone

1. The line or zone separating a grounded glacier from its floating part in the case of **cold** and **polythermal glaciers** (see Figure F.19).
2. The line on the sea floor where a grounded **glacier** terminates in water at an ice cliff, as applied to **temperate** and some **polythermal glaciers**.

Grounding-line fan

A subaquatic fan made up of material emerging from a **subglacial conduit** where a glacier terminates in the sea or a lake. The fan may be exposed at low tide and have some characteristics of a **delta–fan complex** (Figure G.48).

FIGURE G.48   Grounding-line fan at the terminus of Harriman Glacier, Alaska. The fan, lower right, comprises gravel deposited close to sea level from a subglacial conduit (August 1989).

Grounding-zone wedge

A ramp or wedge of sediment deposited by an **ice stream** on a high-latitude continental shelf during temporary still-stands of the **grounding-line** or **zone** during deglaciation. Wedges may reach 100 m in height, tens of kilometres in width and contain tens of cubic kilometres of sediment (Figure G.49). They are commonly associated with mega-scale glaciation lineations. Those that produce a streamlined surface on top of the wedge indicate that the ice remained active during deposition of the wedge of the wedge (Figure G.49).

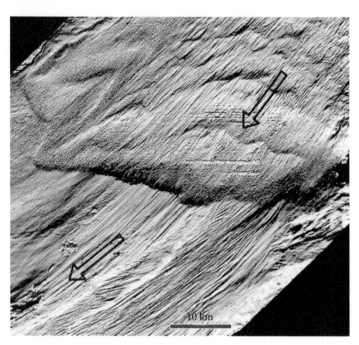

FIGURE G.49 Grounding-zone wedge and mega-scale glacial lineations with arrows showing directions of ice movement in Vestfjorden, north Norway. Using a ship-borne high-resolution multibeam echo-sounders, data covering swaths of up to several kilometres across were obtained. From these data, digital seabed models, accurate to <1 m vertically and between a few metres and a few tens of metres horizontally, were produced. (Image supplied by Julian Dowdeswell.)

Ground moraine

An irregular sheet of glacial sediment (**till**) deposited under a glacier in an area of low relief. It is typically 5–20 m thick, tending to fill hollows and smooth out the topography (Figure G.50). As 'moraine' is generally used for discrete landforms, the term is no longer widely used, and the genetic term **till sheet** is preferable.

FIGURE G.50 Ground moraine, consisting of basal till forms uneven but subdued topography in the proglacial area of Fjallsjökull, Iceland. In the background is Breiðamerkurjökull, the large outlet glacier from Vatnajökull, Iceland's largest ice cap (March 2015).

Growler

A piece of relatively dense glacier ice, floating in the sea and almost awash (Figure G.51). It is smaller than a bergy bit, but because it is difficult to observe, it creates a hazard to ships.

FIGURE G.51   A growler several metres long in the Lazarev Sea, Southern Ocean. The light blue ice is glacier ice, and the dark blue and green veins may represent frozen water-filled crevasses.

Drostke

H

Hanging glacier

A glacier that spills out from a high-level **cirque** or clings to a steep gully or rock face (Figure H.1a and b).

(a)   (b)

FIGURE H.1   Contrasting forms of hanging glaciers. (a) Steep hanging glacier confined to a gully on the Himalayan peak of Ombigaichan, Khumbu district, Nepal (May 2003). (b) A complex of hanging glaciers on the steep mountain face of Aoraki/Mt Cook, South Island, New Zealand (September 2008).

Hanging valley

A tributary valley whose mouth ends abruptly part way up the side of a trunk valley or fjord (Figure H.2), formed as a result of the greater amount of down-cutting of the latter.

H

FIGURE H.2 A hanging valley at the side of Milford Sound, a classic fjord on South Island, New Zealand.

Headwall

The steep ice-draped face at the head of **a valley** or **cirque glacier**. In the case of a valley glacier, the face is prone to **avalanches** (both snow and ice) and rockfall. Glacier tongues below such glaciers are commonly debris-mantled (Figure H.3). In the case of cirque glacier, ice has a more uniform and gentler slope towards the snout; the headwall is exposed when the glacier recedes.

FIGURE H.3 The magnificent 3000-m-high headwall of Lhotse's south face, Khumbu Himal, Nepal, with debris-mantled Lhotse Glacier below (May 2003).

Health of a glacier

An informal term, referring to the **mass-balance** state of a glacier. A healthy glacier has a positive mass balance and this is reflected in a steep, crevassed advancing **snout** (Figure H.4a). An unhealthy glacier has a negative mass balance, and the snout is typically gently graded and smooth (Figure H.4b).

(a)

(b)

FIGURE H.4 Contrasting 'healthy' and 'unhealthy' glaciers. (a) The advancing, heavily crevassed snout of Fox Glacier in April 2008. This was the most recent advance of this glacier, since when it has gone into rapid recession. (b) The low-angle, crevasse-free snout of Midtre Lovénbreen, northwestern Spitsbergen, Svalbard. This glacier has been receding steadily since the end of the 19th century (July 2009).

Highland icefield

A near-continuous stretch of glacier ice, typically tens of kilometres long or wide, but with an irregular surface that follows approximately the contours of the underlying bedrock, and which is punctuated by **nunataks** (Figure H.5).

(a)

(b)

FIGURE H.5   (a) A geologist on skis surveys the scene on Lomonosovfonna, a highland icefield in northeastern Spitsbergen, Svalbard (July 1981). (b) An aerial view of Icefield Ranges, Yukon, Canada, looking towards Mt Logan (left background). This highland icefield, spanning the Alaska–Yukon border, feeds many of the largest glaciers in North America (July 2006).

Hill–hole pair

A close association of a glacier-excavated hollow, commonly with a lake, and a discrete hill with ridges formed by glaciotectonic processes. The hollow may have straight margins that represent strike-slip faults, marking the lateral boundaries of sediment masses thrust out of the depression. The hill comprises the debris that has been displaced and deformed by glacier ice (Figure H.6).

FIGURE H.6 A hill–hole pair near the edge of the Scottish Highlands. Ice flowed east, from right to left, excavating the 'hole' containing the Lake of Menteith, displacing the material to form the part-wooded hill to the left. The depression is mainly less than 15 m deep, whereas the bounding ridge is over 30 m high; the excavated volume approximates that of the hill. The hill is a moraine comprising shelly marine clay, till, sand and gravel, and constructed by glaciotectonic compression. This hill–hole pair was formed during an advance of ice from the southern Grampian Highlands about 12,000 years BP.

Hoarfrost

A deposit of ice produced by condensation of water vapour from the air in sub-zero temperatures. The ice has a crystalline appearance in the form of needles, scales, feathers or fans, adhering to cold surfaces (Figure H.7).

FIGURE H.7 Hoarfrost on branches of a small bush. The ice needles in this case are up to 2 cm long.

Holocene

The current 'epoch' of geological time which is defined internationally as beginning 11,700 years ago, at the end of the last ice age (or **Pleistocene Epoch**). It is broadly equivalent to an interglacial period, but one that has not yet ended. The term *Anthropocene* refers to the period of time, including the present, during which humans have fundamentally altered the planet. It is anticipated that, through international negotiation, a decision will be made in 2016 as to whether *Anthropocene* should be incorporated formally into the geological time scale.

Hook striation

An unusual type of **striation**, whereby a rotating stone embedded in the base of a glacier has scratched a hook-shaped groove alongside normal striations (Figure H.8).

FIGURE H.8   Hook striation and normal striations on basalt bedrock near the margin of Skálafellsjökull, near Höfn, Iceland. Ice flow was from right to left.

Horn

A steep-sided, pyramid-shaped peak, formed as a result of the backward erosion of cirque glaciers on three or more sides (Figure H.9).

FIGURE H.9   The Matterhorn in Switzerland epitomizes the classic horn. Here it is reflected in Riffelsee, near Rotenboden above the village of Zermatt.

Hot-water drilling

The technique of using heaters to boil water and produce steam to drill a hole in a glacier, commonly to the bed (Figure H.10). Typical experiments in the resulting hole include measuring **ablation** and velocity (using stakes), and determining the temperature profile, the state of the bed, **internal deformation** characteristics, water pressure, and allowing for borehole imaging.

H

FIGURE H.10 Hot-water drilling on White Glacier, Axel Heiberg Island, Canadian Arctic in summer 1975. The aim was to insert temperature sensors on a vertical profile down to the glacier bed.

Hummocky moraine

Groups of steep-sided hillocks, several metres high, comprising glacial sediment of both basal and supraglacial origin. Traditionally regarded as the product of ice-stagnation, hummocky moraine also forms at active ice margins where they form aligned belts of debris parallel to the ice margin. **Thrust-faulting** in ice and sediment is implicated in moraine formation in some areas (Figure H.11a). In the latter case, the alternative term **moraine–mound complex** has been used. Hummocky moraines are ice-cored when first formed, but as the ice melts, the moraines may lose their original structurally controlled morphology, and become more subdued (Figure H.11b).

(a)

FIGURE H.11 Recent and Pleistocene hummocky moraine. (a) Fresh hummocky moraine, involving thrusting of basal sediment, produced during the 1948 surge of Kongsvegen, a valley glacier in northwestern Spitsbergen, Svalbard (July 2009). These moraines are still ice-cored and will melt into more subdued forms.

(Continued)

(b)

FIGURE H.11 (Continued)   (b) Hummocky moraine dating from the Younger Dryas cold phase of around 10,000–12,000 years ago, when glaciers were re-established for the last time in Britain, Coire Ardair, Grampian Highlands, Scotland.

Hyaloclastite

Lava (usually basalt) that has become fragmented by quenching in water. In a glacial context this fragmentation takes place as a result of sudden melting when lavas are extruded subglacially (Figure H.12).

FIGURE H.12   Hyaloclastite formed by subglacial eruptions of Pliocene age at Fjordo Bélen, James Ross Island, northeastern Antarctic Peninsula. The dark grey fragments are basalt from the shield volcano of Mt Haddington, which forms the bulk of the island.

Hydraulic jacking/uplift

The effect of pressurised water at the base of a **glacier** lifting the ice. This reduces the area of the **glacier sole** that is in contact with bedrock, thus enabling the glacier to slide more freely. Meltwater and rainwater are present at the glacier surface, and this reaches the bed *via* fractures and **moulins**.

Hydroelectric power generation

Using water from glaciers is an important source of power generation in mountain regions (Figure H.13). The value of glaciers is their ability to store water and deliver it through dry summers. Dams also supply water in the winter when the terrain is frozen or snow-covered.

FIGURE H.13 The hydroelectric reservoir created to capture water from Oberaargletscher in the background. This reservoir forms part of a complex of tunnels and dams that capture water from most glaciers in this part of the Berner Oberland, Switzerland.

Hydrofracturing

The process whereby water under high pressure at the base of a glacier can fracture the underlying sediment. Fracturing may take place along bedding surfaces (Figure H.14), or cut through the strata, allowing sediment to be injected in the form of **clastic dykes**. A second form of hydrofracturing occurs within glacier ice. It is associated with the drainage of supraglacial lakes, and widely known from the Greenland Ice Sheet where large lakes form on the glacier surface and drain to the bed each summer *via* **moulins**.

FIGURE H.14 A thin section of laminated limestone in a glaciotectonised zone below subglacial deposits in the late Precambrian (Neoproterozoic) Wilsonbreen Formation at Ditlovtoppen in northwestern Spitsbergen, Svalbard. The brecciation shown here was the product of hydrofracturing beneath a glacier. These rocks are an integral part of the Snowball Earth theory of global glaciation about 650 million years ago. (Photograph by Edward Fleming.)

I

Ice

The solid form of water formed by the freezing of water, transformation of snow through firn into glacier ice (Figure I.1), the condensation of atmospheric water vapour as crystals in sub-zero temperatures, and the soaking of a porous snow pack by water that subsequently freezes.

FIGURE I.1 A piece of stranded glacier ice, a bergy bit, melting away, on a beach near Ny-Ålesund, Svalbard.

Ice age

A period of time when large ice sheets extend from the polar regions into temperate latitudes (Figure I.2). The term is sometimes used synonymously with **glacial period**, or it embraces several such periods to define a major phase in the Earth's climatic history.

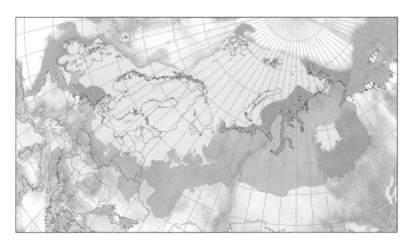

FIGURE I.2 Ice ages over Eurasia, showing the limit of the Last Glacial Maximum or Weichselian glaciation over Eurasia (light blue), together with the maximum extent of ice during the Quaternary Period (dark blue) (Map compiled by Jürgen Ehlers.).

Ice apron (also known as **mountain apron glacier** or **ice wall**)

A steep mass of ice, commonly the source of ice avalanches, that adheres to steep rock near the summits of high peaks (Figure I.3).

FIGURE I.3 A large ice apron, with both smooth and heavily crevassed ice draping the north face of Mt Steele above Steele Glacier, Icefield Ranges, Yukon, Canada (July 2006).

Ice avalanche

The sudden failure of unstable glacier ice results in a rapid descent of fractured ice. Ice avalanches are usually difficult to predict and pose a major hazard to climbers and mountain communities.

Iceberg

A piece of ice of the order of a metre to many kilometres across that has been shed by a glacier terminating in the sea or a lake. There are several types of iceberg (described separately), ranging from **brash ice** at the smallest end of the spectrum, through **growlers** and **bergy bits** of several metres across to **tabular (ice)bergs** and **ice islands** many kilometres across.

Iceberg armada (see **iceberg plume**)

Iceberg crevasses

Large fractures in an iceberg are either inherited from a fast-flowing **ice stream** or formed when an iceberg becomes grounded and new stresses are imposed (Figure I.4).

FIGURE I.4 Two large icebergs with prominent crevasses surrounded by sea ice and small icebergs, Marguerite Bay, Antarctic Peninsula (November 2012).

Iceberg decay

The process of melting, erosion by waves and weathering of icebergs begins soon after they have calved from the parent glacier. Melting takes place when an iceberg becomes grounded (Figure I.5a), and it may disintegrate into ice rubble (**bergy bits**) (Figure I.5b). Large **tabular bergs** may drift far from the coast, decaying slowly (Figure I.5c).

(a)

(b)

FIGURE I.5 Antarctic icebergs in various stages of decay. (a) A grounded, heavily decayed tabular berg off Wormald Ice Piedmont, near Rothera Station (UK), southwest Antarctic Peninsula (December 2012). (b) An iceberg that has rotated several times and undergone partial collapse as it has melted, Rothera Point (December 2012). *(Continued)*

(c)

FIGURE I.5 (Continued)   (c) The remnant of a large tabular berg, with only twin towers surviving above the waterline, close to the Antarctic Convergence in the Southern Ocean (March 1991).

Iceberg graveyard

An area of the ocean where numerous icebergs become grounded and decay *in situ*. Usually graveyards are found in bays (Figure I.6) or on marine banks and shoals.

FIGURE I.6   Iceberg graveyard near Rothera Point, Adelaide Island, southwest Antarctic Peninsula (December 2012).

Iceberg grounding

Grounding of icebergs occurs when icebergs reach shallower waters and become lodged on the sea bottom. They may be released on a rising tide or when they have undergone melting. Tabular bergs, if lodged for many months, gradually tilt and break up (Figure I.7).

FIGURE I.7   Aerial view of grounded tabular bergs breaking up off the coast of Coats Land, Weddell Sea, Antarctica (January 1991).

Iceberg keel

That part of an iceberg lying below the waterline (Figure I.8a). Iceberg keels may form underwater projections that are hazardous to passing vessels (Figure I.8b). Iceberg keels are noted for gouging sea-floor sediments.

(a)

FIGURE I.8   Iceberg keels. (a) The keel of this bergy bit is clearly evident extending down into the clear waters of Prince Gustav Channel on the northeast side of the Antarctic Peninsula.   *(Continued)*

FIGURE I.8 (Continued) (b) Sailing close to this tabular iceberg, underwater projections from the keel are clearly visible. The location is Erebus and Terror Gulf, north of James Ross Island, Antarctic Peninsula.

(b)

Iceberg pits and mounds

Pits are hollows formed on sand or gravel beaches in high latitudes when stranded **bergy bits** become buried and subsequently melt out slowly. Mounds are formed when sediment is first washed over a bergy bit (Figure I.9). These are ephemeral features, barely surviving more than several tidal cycles.

FIGURE I.9 Iceberg pits and mounds on the gravel shore of Engelskbukta, a bay on the northwestern coast of Spitsbergen, Svalbard. The island of Prins Karls Forland is in the background.

Iceberg plough-marks

Formed when **iceberg keels** make contact with and gouge the sea floor or lake bottom. They are commonly seen on beaches with a high tidal range, where sizeable bergy bits are dragged across the beach by longshore drift (Figure I.10). In deeper water, larger icebergs can gouge and churn up large areas of sediment (*cf.* **iceberg turbate**).

FIGURE I.10   Plough-marks made by bergy bits being dragged across a beach by longshore drift and exposed at low tide; Glacier Bay, southeastern Alaska (Summer 1985).

Iceberg plume

Approximately linear zones of icebergs, bergy bits and brash ice that extend from the calving margin of a tidewater glacier or ice shelf (Figure I.11). A rare exceptionally large calving event, which produces numerous large tabular bergs, gives rise to an **iceberg armada**.

FIGURE I.11 An aerial view of an iceberg plume extending from the calving margin of Holgate Glacier, Alaska (July 2011).

Iceberg push ridges

Irregular ridges of gravel or sand, produced when **icebergs** or **bergy bits** become grounded on a beach (Figure I.12). The ridges are exposed at high tide and attain heights of a few metres.

FIGURE I.12 Iceberg push ridge on the shore of Heather Island, Prince William Sound, Alaska. The large number of icebergs is the result of calving during catastrophic recession of Columbia Glacier (Summer 1985).

Iceberg rafting or ice-rafting

The process of transport of debris by icebergs into a lake or the sea. The debris is released and accumulates as **ice-rafted debris** in **glaciolacustrine** (Figure I.13a) or **glaciomarine** environments (Figure I.13), respectively.

(a)

(b)

FIGURE I.13 Iceberg rafting in contrasting environments. (a) A large cluster of debris-covered icebergs calved from Tasman Glacier, South Island, New Zealand, into a rapidly growing proglacial lake. The glacier is just visible in the far left background (April 2007). (b) Scattered angular boulders on a small tabular iceberg close to James Ross Island in the Weddell Sea, Antarctica. This iceberg was probably formed during the catastrophic break-up of Larsen B Ice Shelf (February 2006).

Iceberg scour marks (see iceberg plough-marks)

Iceberg turbate

Sediment disturbed and added to by the impact of drifting icebergs making contact the sea or lake floor.

Ice blister

A swelling of water ice contained within a totally frozen pond on the surface of a glacier or ice shelf (Figure I.14). The top may have a system of radiating fractures. Typically, blisters are several metres in diameter and up to a couple of metres high. They are formed as a result of the expansion of ice in the pond as it freezes.

FIGURE I.14 An ice blister on George VI Ice Shelf, Antarctic Peninsula. Melting produces copious water that collects in depressions in the ice surface, totally freezing each winter. This example is about 1.5 m high and 5 m in diameter (November 2012).

Ice breccia

Coarse angular fragments of broken ice cemented or frozen together into a solid mass (Figure I.15). Ice breccia is typically associated with undercut glacier cliffs in the Polar Regions.

FIGURE I.15 Ice breccia, comprising angular fragments of ice that have fallen from an ice cliff at Fountain Glacier, Bylot Island, Canadian Arctic (July 2014).

Ice cap

A dome-shaped mass of **glacier ice**, largely obscuring the subglacial topography (Figure I.16a and b). Ice caps are usually located over a highland area and are generally defined as covering up to 50,000 km² (in contrast to an **ice sheet** which exceeds this figure). Ice flow is generally radial but is focused within **ice streams**.

(a)

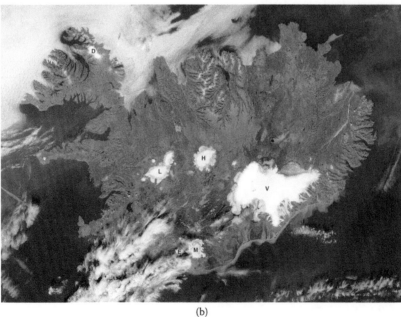

(b)

FIGURE I.16 Polar and temperate ice caps. (a) The elongate cold ice cap on Snow Hill Island, northwestern Weddell Sea, Antarctica, viewed from James Ross Island. A low ice cliff marks its outer edge (January 2006). (b) A satellite image of Iceland, showing the island's temperate ice caps. They are built on volcanic edifices, and subglacial eruptions sometimes occur. The grey outer fringes of these ice caps represent the ablation zones. The largest ice cap is Vatnajökull (V). The smaller ones are Langjökull (L), Hofsjökull (H), Myrdalsjökull (M), Eyjafjallajökull (E) and Drangajökull (D). (Aqua satellite image from NASA taken on 8 August 2004.)

Ice cave

A large cavity that is formed beneath a glacier. Meltwater is usually responsible for eroding out the cave, especially when the glacier is receding and ice deformation is too weak to close it (Figure I.17). Roof collapse is common. Some caves occur where ice detaches from the bed as it flows over a bedrock step.

FIGURE I.17   An ice cave carved out by subglacial meltwater beneath Breiðamerkurjökull, southern Iceland. Note the summer water level where debris has adhered to the ice, level with the feet of the glaciologist (March 2015).

Ice cliff (ice wall)

A vertical face of ice normally formed where a glacier terminates in the sea or is undercut by streams (Figure I.18). See also **ice wall**, which is formed where ice cliffs are grounded on the sea bottom.

FIGURE I.18   Ice cliff at the front of Crusoe Glacier, Axel Heiberg Island, Canada. This section of the cliff is of the order of 20 m high (July 2008).

Ice-contact delta

A delta formed from a subglacial stream where a glacier terminates in the sea or a lake. The top surface is in the intertidal zone (Figure I.19).

FIGURE I.19   Ice-contact delta formed by subglacial meltwater from the tidewater Kronebreen–Kongsvegen glacier complex, northwestern Spitsbergen, Svalbard. The braided streams on the delta top are partly flooded on the rising tide. Note the three geologists on the lower right (August 2001).

Ice-contact fan

A spread of sand and gravel produced from a point source, where a subglacial stream leaves the glacier. The top is braided, and the fan may be abandoned if the glacier recedes and leaves it elevated relative to the surrounding topography (Figure I.20).

FIGURE I.20   An abandoned ice-contact fan in front of the receding Comfortlessbreen, northwestern Spitsbergen, Svalbard. This is a part-tidewater and part-terrestrial glacier. The fan was created from a single subglacial stream, and the ice-contact slope (in shadow) represents an ice position that was stable for several years (July 1992).

Ice core

A cylindrical length of ice extracted from a glacier or ice sheet by mechanical means (Figure I.21). The cores are used to reconstruct climatic records and determine deformation and other attributes of an ice mass. The longest cores are several thousand metres. Those obtained from the Greenland Ice Sheet have provided a record of climatic change spanning over 100,000 years, that is, through the last interglacial–glacial cycle. Cores from Antarctica cover almost the last million years and span several glacial–interglacial cycles. Cores from more temperate mountain regions provide climatic and pollution records typically of a few hundred years (Figure I.21).

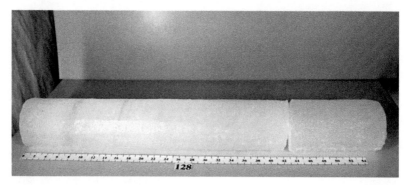

FIGURE I.21   Section of an ice core retrieved in 1994 at 4200 m above sea level on Grenzgletscher, Monte Rosa Massif, Switzerland. The core is from a depth of 45 m and shows a prominent yellow layer associated with a Saharan dust fall in 1977. The drilling project was conducted in order to establish a record of air pollution in central Europe during the 20th century. (Photograph by Anja Eichler.)

Ice-cored moraine

Ridges of debris covering stagnant ice, left behind as a **glacier** recedes. **Lateral moraines** (Figure I.22a) and the **end moraines** (Figure I.22b) of **valley glaciers** commonly have an ice core. The debris cover slows down the melting, and ice-cored moraines may be found several kilometres beyond an active glacier margin.

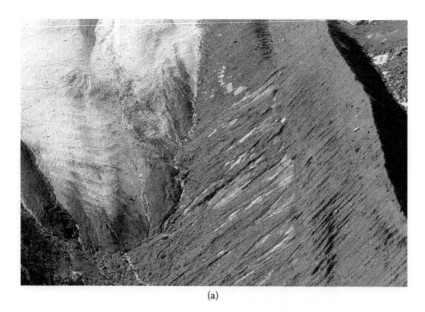

(a)

FIGURE I.22   Ice-cored moraines. (a) Aerial view of an ice-cored lateral moraine at Vadret da Tschierva, Switzerland. The ridge on the right is a Little Ice Age lateral moraine, and the ice core is evident from the white streaks between the ridge crest and the receding glacier margin (July 2004).   *(Continued)*

(b)

FIGURE I.22 (Continued) (b) Ice-cored recessional moraine near the terminus of Kronebreen, northwestern Spitsbergen, Svalbard. The colourful debris cover is thin (<1 m), and the surface topography very uneven. The glacier lies off the picture to the right (July 2009).

Ice crystals

Glacier ice crystals crystallographically are of hexagonal form, but grow from firn into irregular interlocking shapes. Crystals are typically a few centimetres long and contain air bubbles that are sometimes deformed into ellipses (Figure I.23a). Exceptionally, particularly in stagnant ice near a glacier **snout**, individual crystals may exceed 20 cm in length (Figure I.23b). Ice crystals also form on glaciers from the freezing of water in **crevasses**, **moulins** (see **crystal quirks**), supraglacial streams and **cryoconite holes**. Overnight freezing forms thin plates of interlocking ice crystals over these water bodies (Figure I.23c).

(a)

FIGURE I.23 Varieties of ice crystals in glaciers. (a) Typical coarse bubbly ice, here forming a weathered mosaic of interlocking crystals, 3–4 cm in diameter: Stagnation Glacier, Bylot Island, Canadian Arctic.

(Continued)

(b)

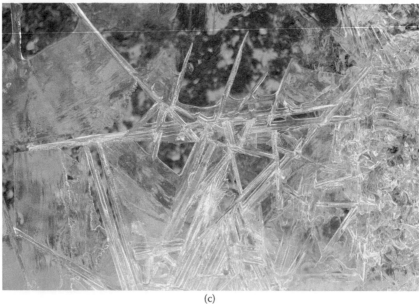

(c)

FIGURE I.23 (Continued) (b) A single large crystal of glacier ice taken from a recently calved iceberg from Columbia Glacier, Prince William Sound, Alaska. (c) A lid of frozen water ice on Vadrec del Forno, Switzerland.

Ice crystal fabric

The arrangement in three dimensions of crystals in an ice mass. Measurement of crystal orientations (typically 50 or more) yields fabric diagrams, which can be related to the stress or strain regimes within a glacier. Crystal fabric also refers to the shape and bubble characteristics of the ice.

Ice crystallography

Examination of the crystal structure of ice, especially glacier ice. This is achieved by creating thin sections, viewing between cross-polaroid plates and measuring the orientation of the crystallographical axis of as many crystals as possible using an instrument called a universal stage (Figure I.24). Alternatively, more efficient methods use automated fabric analysers. Ice crystallography is normally investigated in a cold room at a temperature well below freezing.

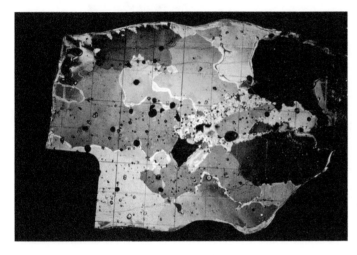

FIGURE I.24 A thin section of glacier ice from Charles Rabotsbreen, northern Norway, viewed between cross-polaroid plates. The grid is 1 cm scale. The thin section was created by melting down a thin block of ice to 1–2 mm thickness on the base of a frying pan, using a Bunsen burner.

Ice-dammed lake

A temporary lake, dammed by a glacier or where two or more glaciers merge. Lakes commonly fill during the summer melt-season, and when full, or when the ice dam is hydraulically lifted, they are prone to catastrophic drainage. Ice-dammed lakes are most common at the margins of polythermal glaciers in the Arctic (Figure I.25a), but many less stable ones also occur in temperate mountain regions, such as the Alps (Figure I.25b).

(a)

FIGURE I.25 (a) The ice-dammed Astro Lake with numerous icebergs and Astro Glacier beyond, Axel Heiberg Island, Canadian Arctic. This lake is dammed by Thompson Glacier in the foreground (Summer 1977). *(Continued)*

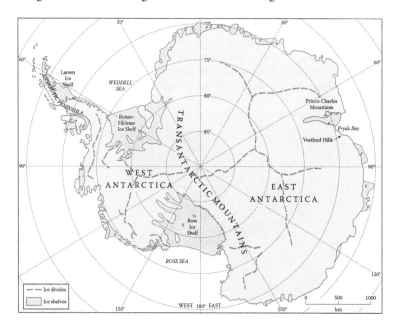

(b)

FIGURE I.25 (Continued)   (b) Gornersee in Switzerland after a massive calving event from Grenzgletscher (left foreground) and Gornergletscher (far right). This ice-dammed lake regularly drains once every summer, sometimes causing floods in nearby Zermatt. The lake has become much smaller in recent years as a result of recession (June 2004).

Ice divide

Analogous to a watershed, an ice divide is the zone of highest elevation in an ice sheet or ice cap from which ice flows in opposite directions (Figure I.26). Ice flow at a divide is near zero, and as such it is possible to find the oldest ice in a glacial system by coring to the bed there. These locations are important for obtaining the longest possible climatic records, as in the Antarctic and Greenland ice sheets. Ice divides do not necessarily coincide with the highest elevations of subglacial bedrock and can migrate over time.

FIGURE I.26   Ice divides, marked by dashed lines, in the Antarctic Ice Sheet. These ice divides define several distinct glacier basins.

Ice dome (see **ice cap**)

Ice facies

The different types of ice in glaciers, based on crystal structure, bubble content and debris distribution. The surface of a glacier has a simple threefold classification of coarse bubbly, coarse clear and fine-grained ice in order of volume (Figure I.27). Ice facies are more complex at the base of a glacier, where basal ice has variable content of debris in addition.

FIGURE I.27 Typical ice facies observable at the surface of a valley glacier, Variegated Glacier, southeastern Alaska. Mid-tone blue is coarse bubbly ice and represents the bulk of the glacier; pale blue is fine-grained ice with the texture of sugar when weathered; and dark blue is coarse clear ice lacking bubbles.

Icefall

The steep reach of a glacier where it flows over a bedrock step. The glacier surface becomes heavily crevassed in the zone of extending flow, making passage on foot difficult and dangerous (Figure I.28).

FIGURE I.28 The icefall of temperate valley glacier, Vadret da Morteratsch, Switzerland. This is more than 200 m high (September 2012).

Icefield (see **highland icefield**)

Ice floe (see **floe**)

Ice flowers (see **frost flowers**)

Icefoot

A narrow fringe of **sea ice** (including frozen sea spray) attached to the coast, unmoved by tidal action, and remaining after the continuous sea ice cover (**fast ice**) has broken free (Figure I.29).

FIGURE I.29 Icefoot near Rothera Point, Adelaide Island, west Antarctic Peninsula. The icefoot cliff stands about 3 m high (November 2012).

Ice fringe

A very narrow **ice piedmont**, with a coastal cliff, extending less than a kilometre from the sea (Figure I.30).

FIGURE I.30 Ice fringe at the foot of steep rock faces, Adelaide Island, west Antarctic Peninsula (October 2012).

Ice front

The vertical cliff forming where an **ice shelf** or other floating glacier meets the sea. The height of the cliff ranges from 2 to over 50 m (Figure I.31). In contrast, an **ice wall** is grounded on the seabed.

FIGURE I.31   Ice front at the northern edge of George VI Ice Shelf, southwest Antarctic Peninsula with Alexander Island in the background. In the foreground is smooth sea ice and scattered icebergs (November 2012).

Icehouse state

One of the two dominant climatic states on the Earth (the other being the Greenhouse state), during which continental-scale ice sheets existed. The term is generally used when discussing Earth's climate on a multi-million year timescale. Icehouse states have existed at least six times since the first extensive ice developed on the planet, about 2700 million years ago. Climate change on this timescale was caused not only by changes in atmospheric composition (especially CO_2) but also by tectonic plate movements and mountain-building, changes in ocean configuration and variations in the Earth's orbit.

Ice island

A **tabular iceberg,** specifically associated with the Arctic. Typical thicknesses are a few tens of metres, and they range in area from a few thousand square metres to as much as 500 km². They originate from floating **glacier tongues** and the few remaining **ice shelves** around Greenland and Ellesmere Island (Figure I.32a). Some ice islands float down through Baffin Bay and occasionally reach as far as Newfoundland, although most disintegrate *en route* (Figure I.32b).

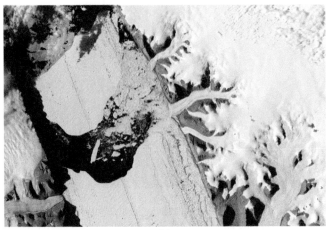

(a)

FIGURE I.32   Ice islands produced by Petermann Glacier, a large outlet glacier from the Greenland Ice Sheet in northwestern Greenland. (a) ASTER satellite image of the 20 km-wide floating terminus of the glacier, northwestern Greenland, in July 2012, showing the birth of an ice island. This large tabular iceberg covered 32 km². (Image from NASA Earth Observatory, petermann_ast_2012203_geo.)   *(Continued)*

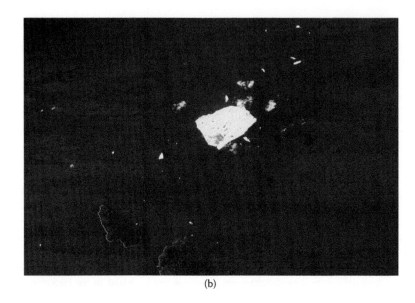

(b)

FIGURE I.32 (Continued)   (b) An earlier ice island produced by the same glacier in June 2010 is here photographed close to the northeastern coast of Newfoundland in August 2011, after travelling several thousand kilometres. Measuring about 4 × 3.5 km, it still carried evidence of the original melt-streams, but is much smaller than its original size. (Image from NASA Earth Observatory, taken with a digital camera from the International Space Station, Astronaut photograph ISS028-E-34749.)

Ice-marginal channel

A channel cut by meltwater at the margin of a valley glacier. Channels are typically steep-sided and several metres deep (Figure I.33). They are associated particularly with polythermal glaciers in the Arctic, where cold ice at the margins does not allow water penetration to the bed.

FIGURE I.33   Ice-marginal channel at Crusoe Glacier, Axel Heiberg Island, Canadian Arctic. The channel is deeply incised into permafrost adjacent to this advancing glacier. Englacial meltwater, which has picked up sediment from the bed, is contributing to flow in the main channel (July 2008).

Ice-marginal lake (see ice-dammed lake)

Ice mélange (see **Sikussaq**)

Ice piedmont

An expanse of glacier ice covering a coastal strip between 1 and 50 km wide. The outer edge is generally a cliff terminating in the sea (Figure I.34).

FIGURE I.34 Ice piedmont descending from Reptile Ridge (right), an arête on Adelaide Island, near Rothera Station (UK), Antarctic Peninsula (November 2012).

Ice-rafted debris (see **iceberg rafting**)

Ice-rafting (see **iceberg rafting**)

Ice rise

An area of elevated grounded ice surrounded by an **ice shelf** or the ice of a floating **glacier tongue**. No bedrock is exposed. Ice rises range in size from a few square kilometres (Figure I.35) to tens of thousands of square kilometres and may attain elevations of several hundred metres. The largest today is Berkner Island in the Filchner–Ronne Ice Shelf, Antarctica. Ice movement on a rise is generally independent of the surrounding ice.

FIGURE I.35 Small ice rises in the northeastern part of George VI Ice Shelf, adjacent to Palmer Land on the Antarctic Peninsula (December 2012).

Ice rumple

A wave-like form in an ice shelf where grounding takes place, but without the elevated nature that character-ises an **ice rise**. Grounding causes buckling of the ice shelf and deflection or stopping of ice flow. Extensive chaotic **crevassing** may occur in such zones (Figure I.36). The long axis of an ice rumple can run for tens of kilometres.

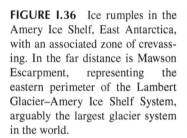

FIGURE I.36 Ice rumples in the Amery Ice Shelf, East Antarctica, with an associated zone of crevass-ing. In the far distance is Mawson Escarpment, representing the eastern perimeter of the Lambert Glacier–Amery Ice Shelf System, arguably the largest glacier system in the world.

Ice sail (see ice ship)

Ice sheet

A mass of ice and snow of considerable thickness and exceeding 50,000 km^2 (Figure I.37a). Ice sheets rest on bedrock or sediment, including regions below sea level, and are fringed by **ice shelves** and **ice walls** (Figure I.37b). The only ice sheets today are the Antarctic Ice Sheet (with three connected component parts: the West, East and Antarctic Peninsula ice sheets) and the Greenland Ice Sheet. Several more ice sheets existed during the Pleistocene ice ages.

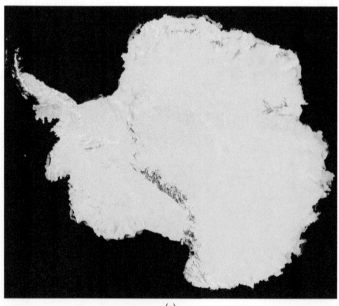

FIGURE I.37 The Antarctic Ice Sheet. (a) High-resolution map of Antarctica constructed from >1100 images of the Landsat 7 satel-lite acquired between 1999 and 2001. Ice covers 98% of the con-tinent; exposed areas are shown as dark brown patches, with the Transantarctic Mountains separat-ing East and West Antarctica being prominent. (Image from NASA Landsat Image Gallery; http://eoimages.gsfc.nasa.gov/images/imagerecords/78000/78592/antarctica_etm_2000001_lrg) *(Continued)*

(a)

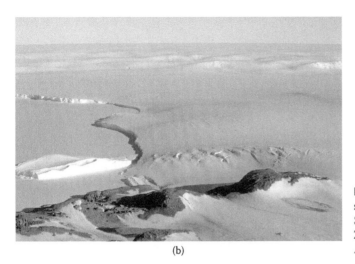

(b)

FIGURE I.37 (Continued) (b) The seaward edge of the Antarctic Ice Sheet, defined by ice walls, near Zhongshan Station (China), East Antarctica (November 1994).

Ice shelf

A floating slab of ice of considerable thickness (up to several hundred metres), fed by ice streams from the interior of an ice sheet, *in situ* snowfall and sometimes by freeze-on of marine water to the base. The seaward edge is termed an **ice front**. The majority of ice shelves are found in Antarctica, the largest (the Ross) reaching an area of almost half a million square kilometres (Figure I.38a and b).

(a)

FIGURE I.38 (a) Edge of an ice shelf between Zhongshan (China) and Davis (Australia) research stations, Princess Elizabeth Land, East Antarctica, February 1995.
(Continued)

(b)

FIGURE I.38 (Continued) (b) Germany's RV *Polarstern* off-loading supplies for Neumayer Station at the edge of Ekström Ice Shelf, Dronning Maud Land, East Antarctica (February 1991).

Ice-shelf collapse

The collapse of ice shelves in Antarctica represents the most dramatic glaciological changes of the present day. A succession of ice shelves in the Antarctic Peninsula has collapsed in recent decades, changes brought about by increased atmospheric and oceanographic warming, inducing increased melt on the surface and on the underside, respectively. Collapse occurs as meltwater weakens the ice along crevasses and other structures, and results in a myriad of small icebergs (Figure I.39a and b).

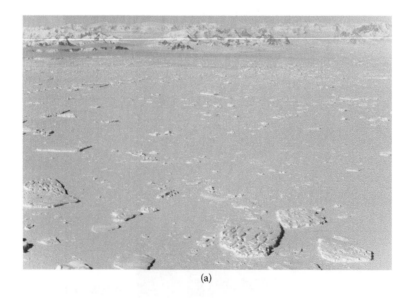

(a)

FIGURE I.39 Aerial and satellite views of ice-shelf collapse in Antarctica. (a) One of the first documented collapses of an ice shelf was that of Wordie Ice Shelf in the southwestern part of the Antarctic Peninsula. This took place between March 2008 and April 2009. This image shows the bay that opened up as a result. In this November 2012 image, many icebergs from the ice shelf remain scattered amongst the winter sea ice.

(Continued)

(b)

FIGURE I.39 (Continued) (b) Satellite image of the collapsing Larsen B Ice Shelf on the eastern side of the Peninsula on 7 March 2002. The pale blue area represents a dense field of overturned icebergs and brash ice. Approximately 3250 km^2 of ice shelf disintegrated in just over a month and produced an iceberg plume that extending out into the Weddell Sea. This is one of a series of images captured by the MODIS instrument on NASA's Terra satellite.

Ice-shelf moraine

Irregular piles, ridges and benches of debris produced at the margins of ice shelves where they impinge upon the coast. Few studies have been made of these features, but at least two processes are involved: (a) deformation of sediment originally entrained as **basal debris** in ice streams supplying the ice shelf (Figure I.40a) and (b) basal freeze on of sediment at the base of an ice-shelf when it temporarily makes contact with the seabed, and then moves to onshore areas (Figure I.40a). Ridges and benches occur at uniform elevations and reach heights of several metres. Typical sediment types are **diamicton** and sandy **gravel**. A few palaeo examples have been reported in the **Quaternary** record from Arctic Canada and Norway.

(a)

FIGURE I.40 Two types of ice shelf moraine in Antarctica. (a) Moraine several metres high at the edge of Alexander Island south of Moutonnée Lake, where ice from the Antarctic Peninsula Ice Sheet has passed through George VI Ice Shelf and impinges on the island. Debris consists of basal glacial material, originating from both local (sedimentary) and distal (igneous and metamorphic) sources (November 2012). *(Continued)*

(b)

FIGURE I.40 (Continued)   (b) Moraine at the edge of the southern McMurdo Ice Shelf, where it impinges on Minna Bluff, a spur of volcanic rock that stretches from the volcano, Mt Discovery, in the background. Here old glacial sediment adheres to the base of the ice shelf, trapped in layers of marine (frazil) ice, that work their way to the surface as a result of net ablation. Movement of ice shorewards concentrates the debris to form ridges and benches of sediment. Scattered boulders of granite, as here, may have been separately deposited from icebergs, before being entrained in the ice shelf (December 2010).

Ice ship/ice sail

A pinnacle of ice, shaped like a triangular sail, typically several metres high, formed as a result of differential ablation under strong solar radiation, usually in low latitudes. They usually form is groups, commonly spanning several square kilometres (Figure I.41).

FIGURE I.41   An extensive field of ice ships at the edge of George VI Ice Shelf, between Moutonnée Lake and Ablation Lake adjacent to Alexander Island, southwestern Antarctic Peninsula (November 2012).

Ice stagnation

The process whereby **dead ice**, often detached from a **glacier** as a result of recession, slowly melts away under a protective cover of debris (Figure I.42). The resulting topography is uneven, with many pools lying between hummocks. See also **hummocky moraine**.

FIGURE I.42 Aerial view of the snout of Pasterzenkees (lower right), in the Hohe Tauern, Austria, with a zone of ice stagnation topography in front (August 2009).

Ice stream

That part of an ice sheet or ice cap in which the ice flows more rapidly, and not necessarily in the same direction as the surrounding ice (Figure I.43a). Strong **crevassing** and zones of strongly sheared ice at the margins of an ice stream (Figure I.43b) are key attributes (*cf.* **shear zone**).

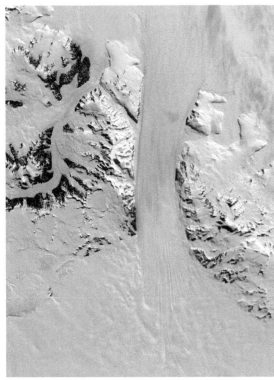

(a)

FIGURE I.43 (a) Landsat 7 satellite image of the massive ice stream of Byrd Glacier, slicing through the Transantarctic Mountains from the East Antarctic Ice Sheet (bottom) to the Ross Ice Shelf. Measuring about 140 km long and 25 km wide, this ice stream arguably drains more ice than any other glacier in the world. The blue areas are zones of net ablation, exposing bare glacier ice. Image from NASA Earth Observatory, taken on 24 December 1999.   *(Continued)*

(b)

FIGURE I.43 (Continued)   (b) Aerial view of a much smaller ice stream near Davis Station (Australia), showing a range of surface characteristics including strong flowlines, crevasses, zones of intense shear, an icefall and small ponds (December 1994).

Ice tongue (see glacier tongue)

Ice tower

Towers occur principally in icefalls in **glaciers** and **ice streams**, where the entire glacier surface is shattered by **crevassing**. Towers are similar to **séracs** but have a particularly large height:width ratio (Figure I.44a). Ice towers also form by differential ablation in areas subject to strong solar radiation. A thin layer of sediment can lower the ice surface more than where the ice is sediment-free (Figure I.44b).

(a)

FIGURE I.44   Two types of ice towers. (a) A temperate glacier with towers in an icefall, Glacier d'Argèntiere, near Chamonix, France (August 2008).   *(Continued)*

(b)

FIGURE I.44 (Continued) (b) Towers resulting from differential ablation of uneven wind-blown sand on the cold Wright Lower Glacier, Victoria Land, Antarctica. The three people to the right provide a scale (January 1999).

Ice vein

Coarse clear ice that is formed in the snow pack or **firn** when percolating meltwater refreezes at an impermeable horizon (Figure I.45). Commonly, this horizon is the base of the previous winter's snowpack. Veins are horizontal or vertical, and usually a few centimetres thick.

FIGURE I.45 Two horizontal ice veins in snow, connected by a vertical 'ice conduit', made more visible by dark cloths. Meltwater percolated from the upper layer to the lower one. The ice veins formed about one year before the image was taken (1980). Ruler length 40 cm.

Ice wall

An **ice cliff** that forms at the seaward margin of an ice sheet or ice cap. It rests on bedrock at or below sea level (Figure I.46a and b).

(a)

(b)

FIGURE I.46   Antarctic ice walls terminating in contrasting bathymetric settings. (a) The ice wall of Barne Glacier terminating in deep water near Scott's Cape Evans Hut, Ross Island. A New Zealand Hägglunds tracked vehicle provides a scale (December 1997). (b) The ice wall of Wormald Piedmont Glacier, near Rothera Station (UK) on Adelaide Island, terminates partly in the intertidal zone (December 2012).

Ice worm

An organism that belongs to the genus *Mesenchytraeus* in the phylum Annelida. These worms spend their lives living in glacier ice and generally retreat below the surface in daylight, since their membranes collapse when exposed to temperatures of more than 5°C. Ice worms can reach several centimetres in length and can be black, brown, blue or white. They feed on algae. First discovered in Alaska, ice worms are found only in the Western Cordillera of North America (Figure I.47).

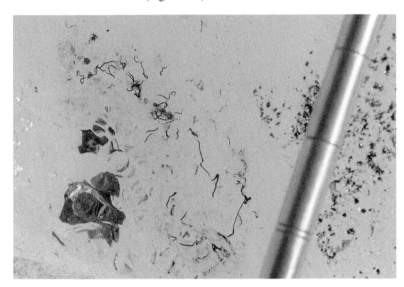

FIGURE I.47   Ice worms on Sholes Glacier, Mount Baker, Washington State, USA. The scale is a probe with 0.1 ft (approximately 2 cm) gradations. (Photograph by Mauri Pelto.)

Icicle

A spike of frozen water resulting from the freezing of dripping water. In warm spells, icicles may break off and collect at the base of the cliff (Figure I.48).

FIGURE I.48   Glaciologist Sean Fitzsimons holds an icicle that he has detached from the ice cliff of Suess Glacier, Victoria Land, Antarctica (February 1999).

Icing (see **Aufeis**)

Imbrication

The alignment of gravel (pebbles, cobbles and boulders) in river deposits, especially glacial streams. The a/b plane dips upstream as a result of backward stacking during high discharge events (Figure I.49).

FIGURE I.49   Imbricated cobbles and boulders in the glacial stream from Franz Josef Glacier, South Island, New Zealand. The glacier lies a few hundred metres to the right of this image.

Incised meander

Sinuous channel cut into the surface of a glacier by a **supraglacial stream** (Figure I.50). Incision results in overhanging sides, and sometimes closure of the upper part of the channel by ice **creep** to give an **englacial conduit**. Deep incision is best seen on **polythermal glaciers**, especially if they are beginning to stagnate.

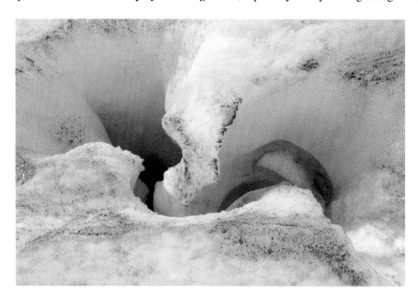

FIGURE I.50   Incised meander, several metres deep, cut into the surface of Gulkana Glacier, southeastern Alaska Range, Alaska (June 2012).

Index glacier (also known as **benchmark glacier**)

A **glacier** that has been selected for long-term investigation, particularly in terms of **mass balance** and hydrology with a view to assessing its response to climatic change. Glaciers chosen are believed to be representative of the areas within which they are located (Figure I.51). Some index glaciers have records spanning over half a century, but globally, relatively few ice masses have been studied.

FIGURE I.51 Gulkana Glacier, southeastern Alaska Range, Alaska 1. This index or benchmark glacier has been monitored continuously by the US Geological Survey since 1966. In 2011, it covered an area of 16.7 km², and had been receding rapidly over this period, and continues to do so (Summer 1989).

Infiltration

In glaciers, infiltration is the entry of water into snow or firn, or is the percolation of water into ice through voids, such as along crystal boundaries. Infiltration is governed by gravity and capillary tension.

Inland ice (from the Danish for *Inlandsis*)

The interior part of an **ice sheet**, commonly fringed by mountains or **nunataks** (Figure I.52). The term refers specifically to the Greenland Ice Sheet.

FIGURE I.52 The eastern edge of the Greenland Ice Sheet or Inlandsis, seen from one of the near-buried mountains, Tillit Nunatak, inland from the coast. The complex of outlet glaciers is flowing left towards the inner reaches of Scoresby Sund (July 1985).

Inter-crevasse blocks

The blocks of ice that form when different sets of **crevasses** intersect. They are often square or rectangular, but degenerate by collapse into **séracs** (Figure I.53).

FIGURE I.53 Aerial view of inter-crevasse blocks in the upper reaches of Fox Glacier, South Island, New Zealand. These blocks occur within the accumulation area, and their sides are several metres in length. The blocks are collapsing into séracs at top left (April 2008).

Interglacial period

A period of time, such as the present day, when ice still covers parts of the Earth's surface, but has receded to the polar or high-mountain regions.

Internal deformation

That component of **glacier flow** which is the result of the deformation of glacier ice under the influence of accumulated snow and firn, and of gravity (Figure I.54) (*cf.* **basal sliding**).

FIGURE I.54 Internal deformation in a small tributary of Aktineq Glacier, Bylot Island, Canadian Arctic. The prominent fold structure has an axis which is broadly parallel to flow (July 2014).

Isoclinal folding

A style of **folding** in ice in which the limbs are parallel to each other. Isoclinal folding is commonly associated with **longitudinal foliation** and originates from earlier layering such as **stratification** (Figure I.55).

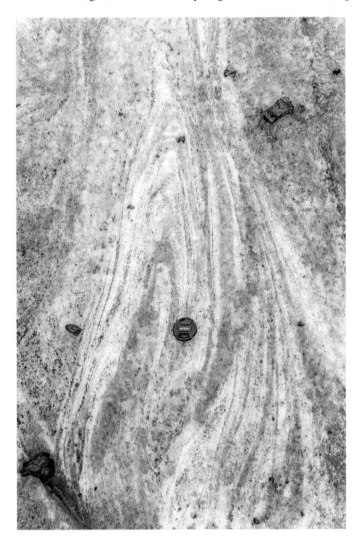

FIGURE I.55   Isoclinal folding associated with longitudinal foliation in Gulkana Glacier, Alaska. Ice flow is from top to bottom of the image; the lens cap is 5 cm in diameter. Ice types represented are coarse bubbly (grey), coarse clear (dark grey) and fine-grained ice (white) (June 2012).

Isostasy (glacio-isostasy)

Crustal deformation resulting from the growth and decay of ice sheets (Figure I.56). An increase in ice mass leads to increased loading of the Earth's crust (**isostatic depression**). Conversely, removal of the ice mass allows the crust to recover (isostatic uplift/rebound). This behaviour arises from the visco-elastic nature of the rock. Glacio-isostasy is a major cause of relative sea-level change. The oceans also have a loading effect on the crust (referred to as **hydro-isostasy**); this also varies according to the behaviour of ice sheets. Maximum loading occurs towards the centre of an ice sheet and is compensated by upward bulging of the crust beyond the margins of the ice sheet to form a **forebulge**. This effect is observed around the edge of the Antarctic Ice Sheet, where inner continental shelf waters are relatively deep and outer shelf waters are relatively shallow. In the Quaternary record, isostatic rebound is illustrated by **raised beaches**.

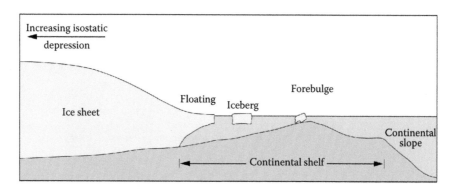

FIGURE I.56 Schematic cross-section through an ice sheet and the adjacent seas, illustrating isostatic loading of the crust and the development of a forebulge on the adjacent continental shelf. (Modified from Figure 2.39 in Lowe, J. and Walker, M., *Reconstructing Quaternary Environments*, 3rd edition, Routledge, London, 2015.)

Isotopes

Varieties of elements, all with identical chemical properties, but not precisely the same physical ones. Specifically, isotopes of the same element have the same number of protons and electrons, but different numbers of neutrons in the nucleus. The difference in the number of neutrons means that isotopes have different mass numbers. In glaciology, especially in ice-core and deep-sea sediment analysis, the ratios of light and heavy isotopes of oxygen are determined (^{16}O and ^{18}O, respectively). This ratio is compared with a universal standard known as Standard Mean Ocean Water (SMOW). The ratio of light to heavy oxygen in snow and ice is related to how much water is locked up in ice sheets and consequently can be used as an historical archive of temperature and past climate (see **ice coring**). Other isotopes commonly used in glaciology and glacial geology are those of hydrogen, deuterium and carbon.

J

Joint-separation

The process of detachment and displacement of large blocks of rock that have fractured along joints and bedding planes (Figure J.1). Joints are exploited by freeze–thaw processes, opening as a result of basal shear stress imposed on the glacier bed. Visible displacements range from a few centimetres to several metres. These blocks may have striated tops.

FIGURE J.1 Detachment of blocks of coarse Cambrian sandstone along vertical joints and low-angle bedding planes in the Rhinogydd (Rhinog Mountains), Snowdonia National Park, Wales. Ice buried these mountains at the Late Glacial Maximum, and ice flow was from east (left) to west (right).

Jökulhlaup (derived from the Icelandic)

A sudden and often catastrophic outburst of water from a glacier during a volcanic eruption (Figure J.2). The term is also used to describe when an **ice-dammed** lake bursts or an internal water pocket escapes, resulting in flooding (*cf.* **glacial lake outburst flood**).

FIGURE J.2 Aerial view over the braided river Markarfljót in southern Iceland during the jökulhlaup caused by the 2010 eruption of the Eyjafjallajökull volcano. Stranded ice debris can be seen on the lower left. (Photograph by Marco Fulle.)

Jökla-mýs (from the Icelandic for 'glacier mice')

Small balls of fluffy moss that form on the surface of a glacier (Figure J.3). They form around sediment or stones, and as the moss grows they protect the ice from ablation, and become elevated on pedestals. With their capacity to hold water, jökla-mýs provide a home for abundant invertebrates, including springtails (**glacier fleas**), tardigrades (water bears) and nematodes (roundworms). Jökla-mýs are best known from Iceland but have also been reported from several other glaciers worldwide.

FIGURE J.3 One of a scattered collection of jökla-mýs growing on the surface of Svinafellsjökull in southern Iceland. The moss ball is roughly 15 cm wide.

K

Kame (from the Scottish *kaim*)

A steep-sided mound or ridge of sand and gravel deposited by supraglacial or ice-marginal streams adjacent to the side or front of a **valley glacier**. Kames usually occur in groups in association with water-filled hollows (**kettle holes**). If mounds are dominant, the term **kame-and-kettle topography** applies (Figure K.1). If, however, kettle holes are dominant the term **pitted outwash** is used. The topography may be either chaotic or have some degree of linearity, related to the manner in which debris concentrations occur in the glacier. This type of topography is typical of receding glacier margins.

FIGURE K.1 An assemblage of mounds (kames) and small lakes (kettle holes) at Fersit near Loch Treig, Grampian Highlands, Scotland. This kame-and-kettle topography was formed during the recession of Younger Dryas glaciers centred on Ben Nevis, around 12,000 years ago.

Kame delta (see **delta moraine** or **grounding-line fan**)

Kame terrace

A flat or gently sloping plain with a steep outer face, deposited by a stream that flows towards or along the margin of a glacier. The terrace is left stranded on the hillside when the ice recedes (Figure K.2).

(a)

(b)

FIGURE K.2 Two examples of kame terraces. (a) A series of kame terraces above the right-lateral margin of Skálafellsjökull, Iceland, from which the ice has receded in recent decades (March 2015). (b) Kame terraces along the flanks of the Tasman River, below Tasman Glacier, South Island, New Zealand. These terraces were formed towards the end of the Late Glacial Maximum.

Katabatic wind (from the Greek *katabatikos,* meaning to flow downhill)

Wind that flows downslope under the influence of gravity. This phenomenon occurs when air is cooled, such as over an **ice sheet** or **glacier**, and becomes denser than air at the same or lower levels. In stable weather conditions, warming of the ground around a glacier during the day is sufficient to create this difference in density and triggers a katabatic wind. These winds typically form in late morning and early afternoon on valley glaciers, but on ice sheets they can flow for many days continuously. Katabatic winds can reach gale or even hurricane force, creating a ground **blizzard**, and they may start and finish without warning (Figure K.3). The pattern of **accumulation** and **ablation** on an ice mass may be controlled to a large extent by katabatic winds.

FIGURE K.3 Clouds of snow (spindrift) herald the arrival of a katabatic wind from the Polar Plateau (East Antarctic Ice Sheet) at Roberts Massif, near the head of Shackleton Glacier. The photograph was taken in calm conditions, just before a gale-force wind arrived at the site (December 1995).

K

Kettle hole (or **kettle**)

A self-contained bowl-shaped depression, commonly containing a pond, within an area covered by glacial sediment (**till**) and **glaciofluvial sediment**. A kettle hole forms from the burial of a detached block of glacier ice by stream sediment and its subsequent melting, or by melting of buried ice in a moraine (Figure K.4) (*cf.* **kame-and-kettle topography**).

FIGURE K.4 A large kettle hole in the outer moraine of Breiðamerkurjökull, the principal outlet glacier from the largest ice cap in Iceland, Vatnajökull (March 2015).

Kinematic wave

The means whereby mass-balance changes are propagated down-glacier. The wave has a constant discharge and moves faster than the ice itself. Kinematic waves are visible as bulges on the ice surface and are especially prominent in surging glaciers. Once the wave reaches the snout, the glacier is able to advance (Figure K.5).

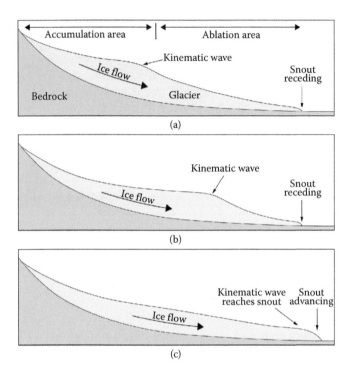

FIGURE K.5 The development and movement of a kinematic wave in a glacier, illustrated by a schematic longitudinal cross-section.

Knock-and-lochain (lochan) topography (derived from the Gaelic)

Rough, ice-abraded, low-level bedrock hillocks with small lakes and boggy areas in the intervening hollows (Figure K.6).

FIGURE K.6 Knock-and-lochain topography at the edge of the East Antarctic Ice Sheet, Vestfold Hills near Davis Station, Australia (November 1994).

L

Lahar

A debris flow consisting primarily of volcanic ash and lava boulders. Heavy rain, and melting snow and ice during a volcanic eruption mix with the loose deposits and form fast-moving tongues of slurry, commonly in existing channels. Lahars are one of the most serious hazards associated with ice-covered volcanoes (Figure L.1).

(a)

(b)

FIGURE L.1 Lahar features on the volcano, Ruapehu on North Island, New Zealand. (a) The channels, along which a devastating lahar flowed on 24 December 1953, following an eruption and partial melting of a glacier in the summit crater. (b) The railway bridge at Tangiwai, which was demolished by the lahar, resulting in destruction of a train and a loss of 151 lives. Both images show the aftermath of a more recent, smaller lahar in March 2007. An early warning system installed after the 1953 disaster led to temporary closure of the railway and adjacent road, thus preventing a further accident.

Lake shorelines (as in **ice-dammed lakes**; see **glacial lake shorelines**)

Landsystem

The classification of the land according to processes, landforms, sediments and soils. A glacial landsystem encompasses a range of depositional and erosional features, which bear witness to the wide range of processes associated with **glaciers: glacial erosion,** and **deposition of glacial, glaciofluvial, glaciofluvial, glaciomarine, aeolian** and **mass-flow sediments.** Defining a glacial landsystem requires detailed geomorphological mapping, documenting the stratigraphy of sediments and assessing which processes have generated each feature. Landsystems defined for modern glacial environments (Table L.1) are the basis for studies of **palaeoglaciology** (former glacier and ice-sheet reconstructions).

TABLE L.1

Principal Types of Modern Glacial Landsystem with Examples

Glacial Landsystem	Well-Studied Examples
Ice-marginal terrestrial landsystem: temperate glacier margins	Iceland
Ice-marginal terrestrial landsystem: polythermal glacier margins	Svalbard, Canadian Arctic, South Georgia
Ice-marginal terrestrial landsystem: cold glacier margins	Dry Valleys, Antarctica
Ice-marginal terrestrial landsystem: polar ice-sheet margins	Vestfold Hills, East Antarctic Ice Sheet
Ice-stream landsystem	Beneath former ice streams in Antarctica and Greenland
Surge-type glacier landsystem	Svalbard, Alaska, Karakorum Mountains
Subaquatic landsystem: fjords	Alaska, Northeast Greenland, Svalbard
Subaquatic landsystem: continental margins	Weddell Sea and Ross Sea, Antarctica
Subaquatic landsystem: proglacial lakes	Receding valley glaciers in Alaska and Iceland
Glaciated valley landsystem	New Zealand (Southern Alps), European Alps, Himalaya, Andes (Cordillera Blanca), Western Cordillera (USA and Canada)
Plateau icefield landsystem	Canadian Arctic, mainland Norway
Paraglacial landsystem	Norway, Karakorum Mountains

Source: Summarised and modified from Evans, D.J.A., 2003, *Glacial Landsystems,* Arnold, London.
Note: There is considerable overlap between these landsystems.

Lateral moraine

An accumulation of debris along the flank of a glacier. It comprises debris of both basal and supraglacial origin, as well as fresh rockfall material. A pair of lateral moraines is most prominent when the glacier recedes, forming sharp-crested ridges with steep furrowed or fluted inner faces and partially vegetated outer slopes. In many mountain regions, the most prominent lateral moraines were formed during the **Little Ice Age** (Figure L.2).

FIGURE L.2 Large lateral moraines dating from the Little Ice Age, supplemented by debris from a 1970s–1980s advance, Vadret da Tschierva, Switzerland (September 2013).

Lateral morainic trough

The name given to the subsidiary valley that forms between the crest of an actively forming **lateral moraine** and the rocky valley side. As the glacier recedes, this valley may be a few hundred metres above the floor of the main valley but may be rapidly destroyed by moraine collapse (Figure L.3). The term **ablation valley** was used in older literature.

FIGURE L.3 The lateral morainic trough along the true right flank of Vadret Pers, Switzerland. The floor of the trough with small lakes is at a considerably higher level than the glacier surface (July 2007).

Latero-terminal (latero-frontal) moraine

An unbroken combination of lateral and terminal moraine, formed by dumping of debris at the margins and front of a **valley glacier** (Figure L.4a and b). This type of moraine may enclose a lake if the glacier has since receded from these limits (see **moraine-dammed lake**).

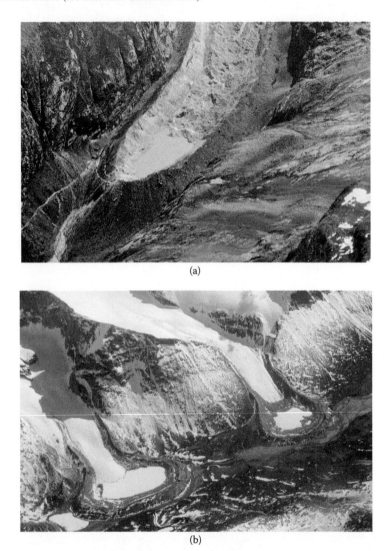

(a)

(b)

FIGURE L.4 Latero-terminal moraines associated with temperate and polythermal glaciers. (a) Debris-mantled Glaciar Llaca, Cordillera Blanca, Peru, with proglacial lake enclosed by moraine. Because of the danger of moraine breaching and catastrophic lake drainage affecting the nearby town of Huaraz, a spillway has been installed to artificially lower the lake (1980). (b) Small, former piedmont glaciers, with ice-cored latero-terminal moraines enclosing frozen lakes, northeastern Baffin Island, Canadian Arctic (July 2014).

Lava-fed delta

A flat-topped, steep-sided volcanic landform constructed by a subaerial basaltic volcano where it enters the sea or a lake (including a glacial lake formed by eruption of a subglacial volcano). A lava-fed delta comprises steeply dipping foresets of disaggregated lava breccia (**hyaloclastite**), capped by flat-lying lava flows (Figure L.5). These two components are separated by a flat-lying internal planar surface, known as a 'passage zone'. Lava-fed deltas range in thickness from a few tens to a few hundred metres.

FIGURE L.5   Lava-fed delta on the north coast of Vega Island, northeastern Antarctic Peninsula. The cliff is over a 100 m high and shows bifurcating steeply dipping foresets of hyaloclastite, representing the frontal part of a delta tongue, capped by a series of basaltic lava flows. A rapidly thinning and melting ice cap lies on top of the delta (March 2012).

Lead (pron. *leed*)

An open crack in sea ice, formed when the ice pack is under tension (Figure L.6). Leads are especially useful for ship navigation and are also zones where marine mammals congregate.

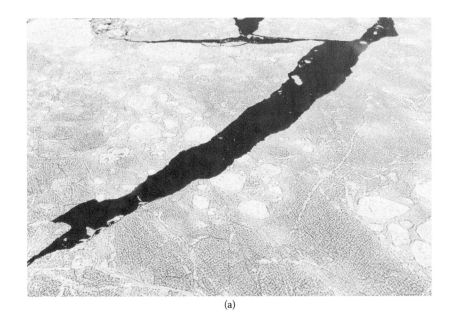

(a)

FIGURE L.6   (a) A lead in sea ice in northern Baffin Bay, developed in the early stages of break-up of winter sea ice.   *(Continued)*

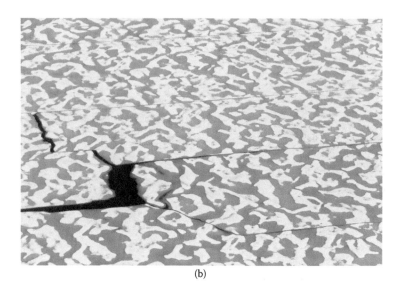

(b)

FIGURE L.6 (Continued)   (b) A lead in rapidly melting sea ice, along which there has been some displace-ment to create a wider patch of open water, Eclipse Sound, Canadian Arctic. The lead slices through sea ice riddled with surface melt ponds (July 2014).

L

Linked cavity system

A network of interconnected cavities that opens up at the bed of a temperate glacier when water pressure at the base exceeds the local ice pressure (Figure L.7). Water at the bed follows these cavities and drainage is efficient, while large parts of the glacier adhere to the bed. The concept was originally developed to explain changes in the hydrological conditions beneath a surging glacier. In such a case, faster ice flow occurs before the cavities develop, when sheet flow and near-complete bed lubrication prevail.

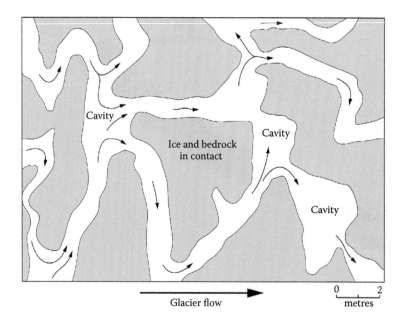

FIGURE L.7   Schematic representation of a linked cavity system, based on the concept by Kamb and others in 1987. Direction of water flow, ice flow and an approximate scale are given for this plan view.

Little Ice Age

A period (also known as the **Neoglacial**) during the **Holocene Epoch** (the last 10,000 years) of expanded valley and cirque **glaciers** worldwide. Several 'Little Ice Ages' have been documented in different regions spanning the period from 1300 to 1900 AD. However, in the Alps and other parts of the world, glacier growth reached a peak between 1700 and 1850 AD (Figure L.8), an event marked by prominent **lateral moraines** and **trimlines**. The gains in glacier mass are attributed to global cooling, resulting in both enhanced **accumulation** and reduced **ablation**.

FIGURE L.8 Mer de Glace (in the foreground) and Glacier de Bossons (far right) near Chamonix, Mont Blanc Massif in 1799 as painted by Antoine Linc. Historical paintings and drawings are important sources for reconstructing the extent of glacier tongues during the Little Ice Age. (Image provided by Heinz Zumbühl.)

Loch

The term given to a lake or an arm of the sea (see also **sea loch**) in Scotland. Most land-locked lochs are elongated and occupy glacially overdeepened basins and have a **parabolic form** in cross-section (Figure L.9).

FIGURE L.9 Loch Shiel in the Western Highlands of Scotland, viewed from above the Glenfinnan monument (in the middle of the photograph).

Lodgement (British) or lodgment (American) (*cf.* **subglacial traction till**)

The process of plastering of glacial sediment onto the bed beneath a moving **glacier** by pressure melting and other mechanical processes. Lodgement occurs in the ice/bed contact zone, especially beneath active wet-based glaciers. Lodgement occurs where the frictional drag between sediment and glacier bed is greater than the shear **stress** imposed by the moving ice, thus causing cessation of further movement of sediment.

The resulting sediment is referred to as **lodgement till** and is a variety of **basal till**. It is typically a poorly sorted sediment with a wide range of **clast** shapes and sizes.

Stones tend to have a preferred orientation parallel to ice flow. Large boulders are prone to being lodged in the surround sediment and then shaped further by glacial erosion (Figure L.10). In recent years, there has been a tendency to move away from the term 'lodgement', as it is now recognised that tills form by a complex interplay of **lodgement**, **melt-out**, bed deformation, flowage and ploughing processes, all of which vary in space and time. Thus subglacial deposits cannot be neatly packaged by each of these individual processes; rather they are a hybrid of some or all of these processes. Instead, the term **subglacial traction till** more realistically covers the range of processes at the ice/bed interface.

FIGURE L.10   A lodged boulder in the proglacial area of Vadret da Morteratsch, Switzerland. The boulder is surrounded by other sediment deposited at the base of the glacier, and traditionally known as lodgement till. The boulder has the shape of a roche moutonnée and indicates ice flow from right to left.

Loess

Wind-blown silt, often derived from the fine-grained material deposited on outwash plains in front of glaciers and carried long distances away from the source (Figure L.11).

(a)

FIGURE L.11   (a) Clouds of silt, whipped up by a strong down-valley gale near Glenorchy, South Island, New Zealand.   *(Continued)*

(b)

FIGURE L.11 (Continued)   (b) Late Holocene loess and interbedded sand exposed in the banks of glacier-fed Matanuska River, Alaska. These deposits are up to 16 metres thick. (Photograph by Helen Roberts.)

Lonestone(s)

A solitary stone of pebble or large size surrounded by finer grained, commonly poorly sorted sediment (Figure L.12). In some cases, lonestones may be dropped into the sediment by **icebergs** (*cf.* **dropstones**), but the process is difficult to demonstrate unequivocally.

FIGURE L.12   A lonestone of dolomite in a matrix of poorly sorted muddy sandstone in the late Precambrian (Neoproterozoic) Wilsonbreen Formation, Ditlovtoppen, northeastern Spitsbergen, Svalbard. This formation is dominated by glacial processes, and in the absence of contrary evidence, this lonestone is interpreted as ice-rafted.

Longitudinal crevasses

Open fractures that are orientated approximately parallel to the lateral margins of a **glacier**. They typically form where the glacier tongue enters a wider part of the valley, or where the glacier emerges from the confines of a valley (Figure L.13). In such cases, the maximum tensile stress is perpendicular to the ice-flow direction. Longitudinal crevasses are similar to **splaying crevasses**.

FIGURE L.13 Longitudinal crevasses near the snout of Rhonegletscher, Valais, Switzerland (September 2009).

Longitudinal foliation

One of two main types of foliation, longitudinal foliation is a pervasive deformation structure found in the majority of **glaciers** (*cf.* **arcuate foliation**). It occurs especially near the margins of a glacier, and where two streams of ice (**flow units**), originating from different accumulation basins, merge. Foliation comprises anastomosing, discontinuous layers of different ice facies, especially coarse clear, coarse bubbly, and fine-grained ice, with individual layers typically a few metres long, but rarely exceeding several centimetres in thickness (Figure L.14a). Foliation is commonly derived from the original layering in the ice (**stratification**), which may have been tightly folded as it transformed into foliation, under a simple **shear** regime. In such cases, the individual layers may be more continuous and extend for tens of metres (Figure L.14b). Foliation sometimes contains disseminated debris, which may either have a basal or supraglacial origin. Differential weathering of the different ice types may give rise to a furrowed surface (Figure L.14a).

FIGURE L.14 Longitudinal foliation in polythermal valley glaciers. (a) Longitudinal foliation on Comfortlessbreen, northwestern Spitsbergen, Svalbard. The furrowed surface results from differential weathering of different ice types (July 1977). *(Continued)*

(a)

(b)

FIGURE L.14 (Continued) (b) Longitudinal foliation with some layers containing debris of supraglacial origin on Stagnation Glacier, Bylot Island, Canadian Arctic (July 2014).

Looped moraine (also referred to as **teardrop moraine** or **surge loops**)

A distinctive type of **medial moraine** on the surface of a surge-type glacier. These moraines are formed of open or closed loops, resulting from tributary glaciers pushing into the main part of a glacier when surging (Figure L.15a and b).

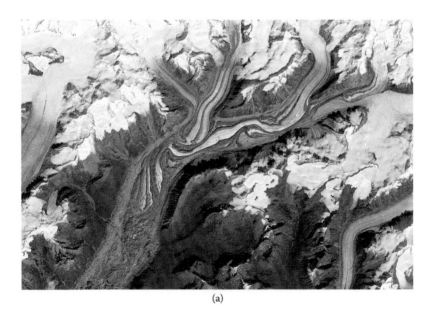

(a)

FIGURE L.15 Glaciers with a history of multiple surges, with several looped moraines during the quiescent stage in the Alaska Range, Alaska. (a) An ASTER satellite image of Susitna Glacier, Alaska, 27 August 2009. This false-colour image shows debris-covered ice as brown and vegetation as red. Available at http://eoimages. gsfc.nasa.gov/images/imagerecords/46000/46449/susitna_ast_2009239_lrg. *(Continued)*

(b)

FIGURE L.15 (Continued) (b) Oblique aerial view of looped moraine in Muldrow Glacier, which flows off Denali (formerly Mt McKinley), the mountain in the background (June 2012).

Low-angle crevasses

Crevasses that form in a near-horizontal attitude above a cavity beneath a glacier, such as in a glacier portal (Figure L.16). Erosion by meltwater induces tensile stresses perpendicular to the roof of the cavity, resulting in unpredictable collapse of the roof.

FIGURE L.16 Low-angle crevasses in the roof of the glacier portal at Franz Josef Glacier, South Island, New Zealand in April 2008. Tourists entering these caves have been killed or injured as a result of roof collapse.

M

Marginal crevasses

Open fractures that extend from the glacier margin towards the middle of a **valley glacier** (Figure M.1), and form in a zone of simple **shear**, associated with drag along the valley sides. **Chevron crevasses** are a particular type of regularly spaced straight marginal crevasses which point up-glacier, typically at 45° to the margins.

FIGURE M.1   Marginal crevasses in lower Grosser Aletschgletscher, Switzerland. Note how these cut across the near medial moraine and ingest debris (July 2008).

Mass balance (or **mass budget**)

The change in mass of a **glacier** over a stated period of time, usually a year or a season (winter or summer). Mass balance is a measure of the state of **health** of a glacier, reflecting the balance between accumulation and ablation, and the position of the **equilibrium line** (Figure M.2). A glacier with a positive mass balance in a particular year gains more mass through accumulation than is lost through ablation; the reverse is true for negative mass balance. Traditionally, mass balance has been measured using stakes inserted into the surface of a valley glacier to record elevation changes. Satellites now document surface-elevation changes of the larger ice masses.

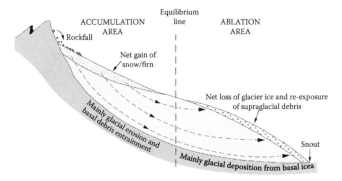

FIGURE M.2   Schematic longitudinal cross-section of a glacier to illustrate ablation and accumulation areas and the position of the equilibrium line.

Meander

A winding curve or bend in a single-channel river or stream. The geomorphological and physical attributes of meandering rivers have been well studied, but outflow from glaciers tends to produce **braided rivers** instead. However, meandering streams are common on glacier surfaces. Such supraglacial meanders grow from relatively straight stream courses that may be controlled by ice structure to highly sinuous forms over several years (Figure M.3). Meanders are typically deeply incised with overhanging walls and, as incision proceeds, the upper walls of the channel may close as a result of ice **creep**. See also **supraglacial streams**.

FIGURE M.3 Extremely well-developed meanders on the surface of a polythermal valley glacier: Vibeke Gletscher, Northeast Greenland National Park (July 1988).

M

Medial moraine

A linear zone of debris within the ablation area of a **valley glacier** system, formed by the convergence of two separate tributaries. The debris comprises sediment of both **basal** and **supraglacial** origin. After convergence basal debris becomes englacial and extends down through the combined glacier. As this englacial debris melts out, the moraine grows larger down-glacier, commonly forming a ridge (Figure M.4a and b). Although normally parallel to ice flow, this relationship may be altered by encroachment of ice into a side valley, or by surging. Medial moraines may also emerge from ice within the ablation area, where they can be traced back to rockfalls in the headwall, that are subsequently folded along with stratification, emerging as the hinges of tight folds.

FIGURE M.4 Medial moraines in temperate valley glaciers. (a) Relatively straight moraines in Kahiltna Glacier, Alaska Range, Alaska, with Denali in the background (July 2011). *(Continued)*

(a)

(b)

FIGURE M.4 (Continued)   (b) The sinuous trough containing Kaskawulsh Glacier, Icefield Ranges, Yukon, generates a fine set of curving moraines (July 2006).

Megablock (see **raft**)

Mega-scale glacial lineations

M

Strongly elongate ridges of sediment, similar to, but much more elongate than, **drumlins**. The land or continental shelves on which they occur is relatively flat and the lineations give the landscape a corrugated surface (Figure M.5). Individual features are typically several tens of kilometres and several hundred

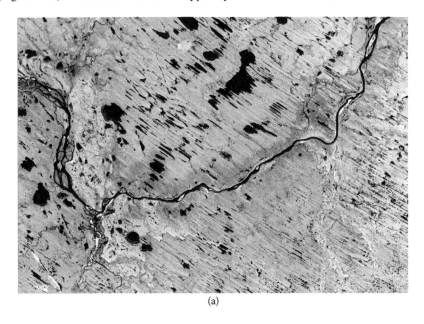

(a)

FIGURE M.5   (a) Satellite image of mega-scale glacial lineations formed under the Laurentide Ice Sheet in the Dubawnt area, Nunavut, Canada. Indiviidual ridges are up to 20 km in length. (Landsat image processed and supplied by Chris Clarke).   *(Continued)*

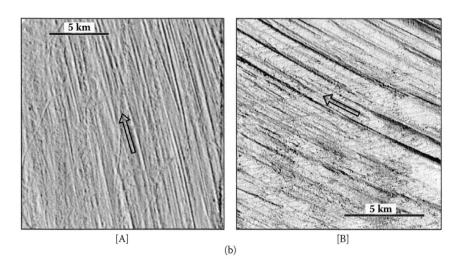

[A]   [B]

(b)

FIGURE M.5 (Continued)   (b) Mega-scale glacial lineations on high-latitude continental shelves. (A) From the outer part of Marguerite Trough, Antarctic Peninsula. (B) From Traena Trough, north Norway. Arrows show the former directions of ice flow. These images were obtained using high-resolution multibeam echo-sounders from ships. (Images provided by Julian Dowdeswell).

metres wide, with spacings up to 5 km. Terrestrial examples are not usually clearly visible at ground level but are well seen in aerial photographs or satellite images. Offshore images are obtained using echosounders. The lineations are interpreted as the product of glacial deposition during fast **glacier flow**, such as beneath **ice streams** (Figure M.5b).

Mélange or ice mélange (see *Sikussak*)

Melt-out

The process whereby embedded debris within a **glacier** melts out at the base or at the surface (Figure M.6). The debris that accumulates from basal melting is referred to as **melt-out till** and is one of the components of **subglacial traction till**. Surface melt-out yields supraglacial till (see **till**).

FIGURE M.6   The melting-out process at the margin of Vadret da Morteratsch, Switzerland, September 2014. Note the accumulated debris, which is melt-out till, in the lower half of the photograph.

Meltwater channel

A channel carved out of sediment and bedrock by glacial meltwater. Lateral meltwater channels are associated with polythermal glaciers, and are steep-sided and deep relative to their width (Figure M.7). See also **glacial spillway** and **subglacial meltwater channel**.

FIGURE M.7 Meltwater channel along the right-lateral margin of Trapridge Glacier, Yukon, Canada. This view, looking upstream, shows how the stream from this polythermal glacier has cut into weathered bedrock, as well as unconsolidated sediment (July 2006).

Microbial processes

Glaciers provide unique ecosystems for cold-tolerant organisms including animals (see **glacier fleas**, **ice worms**), plants, fungi, protists and bacteria. Microbial processes are variable, depending on whether the living medium is snow, ice, debris or water. Key factors in microbial processes are the availability of sunlight, nutrients and stability of the substrate. Cryoconite holes and snow are especially favourable for the growth of microorganisms (Figure M.8a). Sediment provides the nutrients for rich assemblages of cyanobacteria,

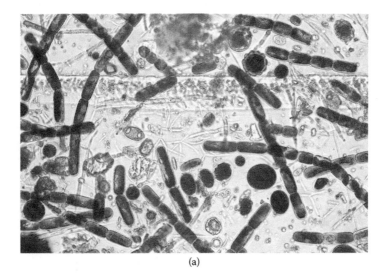

(a)

FIGURE M.8   Examples of microbial processes in glacial environments. (a) Microbial bloom from the snow-pack of Storglaciären near Tarfala, Arctic Sweden, in August 2014, viewed under the microscope at 100× magnification. The image width is approximately 500 μm, and the organisms visible are highly pigmented brown/purple ice and snow algae (red spheres), fungal spores (the spiky structures) and a variety of protozoa and other microbes (Image from Arwyn Edwards and Ottavia Cavalli).   *(Continued)*

M

(b)

FIGURE M.8 (Continued) (b) Algal mounds (stromatolites) growing from the bed of an ice-contact lake and breaching the permanent lid of ice, Lake Brownworth, Wright Lower Glacier, Victoria Land, Antarctica. The lens cap is 5 cm in diameter (January 1999).

green algae and fungi. Frozen ice-contact lakes in Antarctic (e.g. in the Dry Valleys of Victoria Land) also provide a home for living organisms, where algae in particular grow into distinctive mounds known as **stromatolites** (Figure M.8b). Microbial processes are also known from subglacial environments. In a subglacial lake beneath the West Antarctic Ice Sheet microbes are implicated in methane production.

Mirage

An optical phenomenon in the atmosphere in which rays of light are refracted to produce a displaced image of distant features in the sky. Mirages are common in the polar seas, where distant **icebergs** (Figure M.9) and land masses may appear suspended in air.

FIGURE M.9   Mirage of tabular icebergs at sunset, seen from Australia's Davis Station, East Antarctica (March 1995).

Moat

A linear zone of water between **sea ice** and a grounded or floating glacier margin, resulting from opening of a gap between them (Figure M.10).

FIGURE M.10 A moat between Bowers Piedmont Glacier and late winter sea ice (foreground) on the coast of Victoria Land, Antarctica. Ross Island, with the volcano Mt Erebus, are in the background (October 1987).

Morainal bank

A bank formed below water-level in front of a stable grounded glacier terminus in a fjord or lake (Figure M.11a). A bank is rarely well exposed above water level, so is not easy to identify. The bank comprises a locally deformed mix of sediments that are the result of basal glacial deposition, dumping, push and remobilisation of sediment (Figure M.11b).

M

(a)

FIGURE M.11 (a) Morainal bank close to McBride Glacier, Glacier Bay, Alaska. In this low-tide photograph, large bergy bits are stranded on the gravel surface as a consequence of the high tidal range (June 1986).
(Continued)

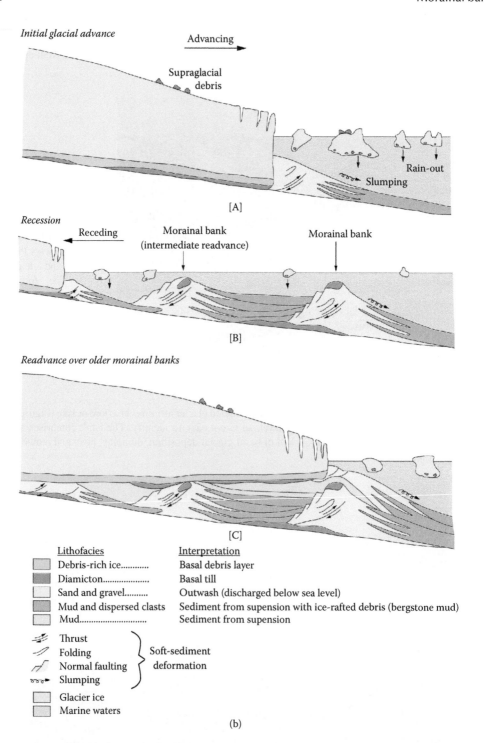

FIGURE M.11 (Continued) (b) Schematic diagram of a cross-section through a tidewater glacier, illustrating the formation of two morainal banks, and modification of them by a readvance. This diagram is based on the sedimentary record recovered by the Cape Roberts Drilling Project in western McMurdo Sound, Antarctica. (From Hambrey, M.J., Barrett, P.J. and Powell, R.D. 2000. In: *Glacier-Influenced Sedimentation on High-Latitude Continental Margins* (eds.) Dowdeswell, J.A. and Ó Cofaigh, C., Geological Society London Special Publication 203, 115–129.)

Moraine

Moraines are landforms that occur both on valley glaciers and in all marginal areas of glaciers and ice sheets. There is a wide terminology to cover all different types of moraines, each of which is described and illustrated separately in this volume, with the general context being given below. On the glacier itself, the principal forms are lateral moraines, and medial moraines that are a combination of two laterals. A variant of these is the **looped or teardrop moraine**, which is a medial moraine distorted during a glacier surge.

Moraines beyond the ice margin are distinct ridges or mounds of debris, laid down directly by a glacier or pushed up by it. These moraines are classified according to position, morphology and process of formation. They consist of subglacial and supraglacial debris, as well as fluvial, lake or marine sediments. Longitudinal moraines (parallel to the valley) include a pair of **lateral moraines** that form along the valley sides, and **fluted moraine** which forms a series of ridges beneath the ice, parallel to flow.

Where supraglacial debris at the glacier snout is deposited on subglacial sediment and mixed with other sediments during a stationary phase, and then subject to reworking by mass-flowage, **dump moraines** are formed. Transverse (cross-valley) moraines include a **terminal** (or **end**) **moraine** which forms at the furthest limit reached by the ice, and **recessional moraines** which represents stationary phases during otherwise general retreat. A **push moraine** is a more complex form that develops especially in front of a cold or poly-thermal glacier during a period of advance. This category includes **annual push moraines,** which form as a result of winter readvances, and **thrust** or **thrust-block** moraines which relate to glaciotectonic processes. Other moraines that are controlled by the structural disposition of debris within the ice are called **controlled moraines**. Disorganised mounds of glacially deposited debris are referred to as **hummocky moraines**, with the connotation that ice-wastage is the dominant process. Similar to these is the **moraine–mound complex,** which comprises organised mounds with a clear geometrical relationship with the glacier margin.

Where continuity exists between a lateral and terminal moraine, the term **latero-terminal moraine** applies. Morainic debris (basal glacial sediment) that is spread across the landscape, smoothing out its irregularities is **ground moraine**. **Breach-lobe moraines** are subsidiary moraines when a glacier readvances and breaches an older moraine to create a lobate form. Moraines associated with marine environments include **ice-shelf moraines** and **De Geer moraines**.

Other moraines, which focus on specific attributes include **boulder moraines**, associated with cold-ice transport, **furrowed (lateral) moraines** that are the result of gullying and **ice-cored moraines** which can be any of the above.

Moraine breach

Occurs where a supraglacial stream cuts through a lateral or terminal moraine and flows away from the glacier (Figure M.12a), or where side-valley streams cut through the moraine to enter the glacier environment (Figure M.12b).

(a)

FIGURE M.12 Two types of moraine breaches. (a) Supraglacial streams collect on Khumbu Glacier, Nepal, to form a single breach in the latero-terminal moraine (April 2003). *(Continued)*

M

(b)

FIGURE M.12 (Continued)   (b) River from Murchison Glacier has cut through the lateral moraine of Tasman Glacier to flow into its proglacial lake, forming a small delta. Previously Tasman Glacier filled the lake basin, and the Murchison River flowed outside the lateral moraine (April 2008).

Moraine-dammed lake

A lake formed as a glacier recedes from its terminal moraine, the moraine acting as a dam (Figure M.13a). Alternatively, where a side glacier with a large latero-terminal moraine enters the main valley, a lake may be impounded (Figure M.13b). Some moraine dams are ice-cored and are thus potentially unstable (see also **glacial lake outburst flood**).

(a)

FIGURE M.13   Two well-known moraine-dammed lakes. (a) Imja Tsho is a rapidly growing supraglacial lake in the Khumbu Himal, Nepal. It is dammed by a wide terminal moraine, the inner part of which is ice-cored (May 2003).   *(Continued)*

(b)

FIGURE M.13 (Continued)   (b) Laguna Parón in the Cordillera Blanca, Peru, is dammed by the debris-covered glacier and its latero-terminal moraine, which descends from the right (1980).

Moraine–mound complex

An assemblage of small mounds of glacial debris. The mounds are typically several metres high, sometimes form in linear belt and have an ice core (Figure M.14). The term is an alternative to **hummocky moraine**, which has connotations of **ice stagnation**. The term embraces various processes, including glaciotectonically controlled deposition, deposition at a dynamically active ice margin, as well as ice stagnation.

M

FIGURE M.14   Moraine–mound complex at Austre Lovénbreen, northwestern Spitsbergen, Svalbard, of Neoglacial age (c. 1900) and younger. This example shows uniform slopes dipping towards the glacier (off the picture to the left) and is interpreted as the product of englacial thrusting when the glacier was over the site (July 2009).

Moulin (from the French for mill)

A water-worn pothole formed where a surface melt-stream exploits a weakness in the ice (Figure M.15). Many moulins are cylindrical, several metres across and extend down to the glacier bed, although often in a series of steps.

M

FIGURE M.15   A supraglacial stream entering a moulin on Vadrec del Forno, Switzerland. The stream has exploited a fracture (crevasse trace) in the glacier surface. The moulin is about 3 m in diameter (August 2011).

Mountain-foot glacier (also known as **mountain-apron glacier**)

A glacier of irregular outline, elongate along slope, formed near the foot of a mountain range (Figure M.16).

FIGURE M.16   Mountain-foot glaciers in the Transantarctic Mountains near Shackleton Glacier, Antarctica. Some of these glaciers adjoin the Ross Ice Shelf to the right. The blue-ice areas are zones of net ablation, generated by strong winds from the mountains (December 1995).

Mountain glacier

A glacier that is confined by mountainous topography and having its own accumulation area (Figure M.17) (*cf.* **valley glacier**).

FIGURE M.17 A mountain glacier in the heavily ice-covered Transantarctic Mountains. Despite proximity to the East Antarctic Ice Sheet, this glacier is self-contained until it merges with the Ross Ice Shelf at its foot (February 1996).

Mud

Sedimentary particles finer grained than sand (Figure M.18). The term is often used loosely; furthermore, usage in sedimentology is inconsistent. The most useful definition in glacial geology is for mud to equal silt + clay according to the Wentworth/Krumbein particle-size scale. Mud therefore is finer than 62.5 μm or 1/16 mm. It represents the fine-grained end member of the threefold classification of particle size in poorly sorted sediments: mud, **sand** and **gravel**. Mud represents the products of glacial abrasion and is commonly found in basally deposited glacial sediments.

M

FIGURE M.18 A patch of mud that has been winnowed out of a debris flow originally deposited as subglacial till, with the glacier Kongsvegen, northwestern Spitsbergen, in the background. Polar bear footprints for scale (August 1995).

Mudflow (see debris flow)

Used loosely, this term embraces saturated flows of sediment irrespective of particle-size distribution. More accurately, the term can be applied to flows consisting almost entirely of mud (silt + clay) particle size (Figure M.19).

FIGURE M.19 Mudflow flowing off debris-bearing ice onto avalanche snow on Gaißbergferner, Tirol, Austria. Each lobe of sediment ceases to flow as the snow absorbs the water from each pulse of saturated sediment (September 2014).

Multi-year ice

Sea ice that has grown over successive winters in the polar oceans. It is especially common in the Arctic Ocean, from which individual **floes** escape and become stranded on beaches on land surrounding the ocean. Multi-year ice is several metres thick and when packed into embayments can be a hazard to shipping (Figure M.20).

M

FIGURE M.20   Multi-year ice floes, several metres thick, transported out of the Arctic Ocean by currents, and stranded on the island of Nordaustlandet, Svalbard (July 1978).

N

Nail-head striae (see **striations**)

Névé (derived from the French)

(a) A synonym of **firn** but a less well-used term today. (b) A snowfield, roughly equating with the accumulation area of a glacier but not precisely defined (Figure N.1).

FIGURE N.1 The névé region, corresponding approximately to the accumulation area, of Fox Glacier, South Island, New Zealand (April 2008).

Niche glacier

A small elongate glacier that occupies a gully or an inclined depression (Figure N.2).

FIGURE N.2 Three niche glaciers occupying gullies on an unnamed mountain near Fountain Glacier, Bylot Island, Canadian Arctic (July 2014).

Nunatak (from the Inuit)

An island of bedrock or a mountain projecting above the surface of an ice sheet or highland icefield (Figure N.3).

FIGURE N.3   Nunataks projecting through the Harding Icefield, Seward Peninsula, Alaska (July 2011).

Nye channel or N-channel (named after the British physicist, John Nye)

A channel cut into bedrock by subglacial meltwater under high pressure. The channel is usually less than 1 m wide, and commonly deeper than it is wide (Figure N.4).

N

(a)   (b)

FIGURE N.4   Single and multiple Nye channels in Pleistocene and contemporary settings. (a) Channel cut into basalt on the southern shore of Loch na Keal, Isle of Mull, Scotland. This, and associated features was formed in the boundary zone of local Mull ice and SW-flowing mainland ice. (b) Multiple channels cut in limestone amongst streamlined bedforms near the margin of Glacier de Tsanfleuron, Switzerland.

O

Oasis (Antarctic oasis)

A large area free of snow and ice in the 98% ice-covered continent of Antarctica (Figure O.1). 'Desert' conditions prevail because of (a) the lack of precipitation, (b) strong solar radiation which melts any snow and ice that forms (except on lakes) and (c) strong katabatic winds that facilitate snow-removal and sublimation.

FIGURE O.1   A view of one of the larger oases in East Antarctica, Amery Oasis, from a high vantage point. Radok Lake in this image is frozen and surrounded by bedrock and glacial sediments spanning many millions of years. The oasis is surrounded by ice, including the Amery Ice Shelf (February 1995).

Obliquity of the ecliptic

Variations in the Earth's axis of tilt between 21°39′ and 24°36′, which takes place on a 41,000-year cycle. This is one of three astronomical variables that affect the amount of solar radiation received at the Earth's surface and is embraced by the Astronomical Theory of Climate Change.

Ogives

Arcuate waves or bands, with their apices pointing down-glacier, which develop in an icefall. **Wave ogives** form first and comprise low-amplitude arcuate waves with a wavelength of tens of metres (Figure O.2a). They reflect differences in ice thickness as ice passes through the icefall – summer melt creates the troughs, and winter accumulation or lack of melting produces the high areas. Wave ogives commonly pass into **band**

ogives (also known as **Forbes' bands** after 19th-century glaciologist, James Forbes), which are arcuate light and dark stripes crossing a glacier surface (Figure O.2b). They consist of intercalated layers of coarse bubbly, coarse clear and fine-grained ice. The lighter layers are predominantly of coarse bubbly ice derived during transformation of snow to ice. The darker layers contain mainly coarse clear ice derived from closed **crevasses** and related veins. Together, they are a type of **arcuate foliation** produced by longitudinal compression and shearing extending from the bed at the base of the icefall. A pair of waves and light and dark bands represent one year's movement through an icefall in a dynamically active glacier. Their arcuate nature becomes more accentuated as the ice flows downhill as a result of faster flow in mid-glacier.

(a)

(b)

FIGURE O.2 (a) Wave ogives on Gates Glacier, a tributary of Kennicott Glacier in the Wrangell Mountains, Alaska (July 2011). (b) Band ogives or Forbes' bands on Mer de Glace, the longest valley glacier in the Mont Blanc region of France. The ogives are generated in an icefall between roughly 2700 and 2500 m above sea level and were first studied by Forbes in the early 19th century (August 2008).

Outlet glacier

A valley glacier that drains an ice sheet (Figure O.3a), a **highland icefield** or an **ice cap** (Figure O.3b). The glacier basin may not be well defined in the accumulation area as it may connect with the heads of other outlet glaciers.

(a)

(b)

FIGURE O.3   Large and small outlet glaciers. (a) A major outlet from the East Antarctic Ice Sheet, Liv Glacier flows for approximately 80 km from the Polar Plateau to the Ross Ice Shelf (February 1996). (b) The short outlet glacier, Middalen, from the ice cap of Hardangerjøkul, near Finse on the Bergen–Oslo railway (August 2008).

Outwash plain

A flat spread of debris deposited by meltwater streams emanating from a glacier or ice sheet (see **braided river**; Figure B.25).

Overdeepened basin (or **overdeepening**)

A deep water-filled hollow eroded by a valley glacier primarily in bedrock (Figure O.4a and b). It occurs inside the **end moraine** complex of the **glacier** and may be substantially deeper than all adjacent areas, including offshore on the continental shelf. Analogous features are a **rock basin** and **fjord.**

(a)

(b)

FIGURE O.4   (a) Overdeepened basin in front of rapidly receding and calving Fjallsjökull, southern Iceland, taken from the last recessional moraine to develop prior to lake expansion (March 2015). (b) Late Pleistocene overdeepened basin, Loch Avon, Cairngorm Mountains, Scotland.

Overspill or **overflow channel** (see **spillway**)

Oxygen isotope analysis

The technique of examining the ratios of the heavy and light isotopes of oxygen in snow, **ice cores** or marine **sediment cores** compared to a standard (SMOW – Standard Mean Ocean Water). Isotopic trends can indirectly provide a climatic record as well as an estimate of the volume of glacier ice on the Earth.

P

P-forms (or plastically moulded forms)

Smooth rounded forms of various types cut into bedrock by the combined erosive power of ice, meltwater and subglacial sediment under high pressure (Figure P.1).

FIGURE P.1 Glacially smoothed bedrock displaying a variety of p-forms, including channels, that are the product of both glacial abrasion and subglacial water erosion, Columbia Glacier, Prince William Sound, Alaska (July 1989).

Pack ice

An area of sea, lake or river ice (excluding **fast ice**), irrespective of how it is disposed. Pack ice cover is normally described by mariners in tenths. Alternatively, descriptive terms are used, for example 'open' or 'close' pack ice (Figure P.2).

FIGURE P.2 Close pack ice with smooth floes and pressure ridges, surrounding icebergs near Mawson Station (Australia), East Antarctica (March 1995).

Palaeoglaciology

The study of glacial geology and geomorphology, based on modern understanding of glacial processes and the spatial arrangement of landforms and sediments (**landsystems**), with a view to reconstructing former glaciers and ice sheets.

Pancake ice

Pieces of freshly formed **sea ice** or lake ice, approximately circular, about 0.3 to 3 m in diameter. The pieces have raised rims, resulting from collision with each other (Figure P.3).

FIGURE P.3 Pancake ice near Davis Station (Australia), East Antarctica (March 1995).

P

Parabolic form

Refers to the shape of a glacial valley or col in cross-section. A parabola is a symmetrical curve and is defined by a quadratic equation of the form $y = ax^2 + bx + c$. **U-shaped valleys** are represented by parabolas with near, vertical sides, but most glacial valleys have less steep sides (Figure P.4).

FIGURE P.4 Two parabolic cross-valley profiles viewed from Loch Etive, Grampian Highlands, Scotland. The nearer open profile, just beyond the loch is at the mouth of Glen Etive, a major glacial trough modified most recently about 12,000 years ago during the Younger Dryas. In the background, the steeper parabolic cross-section is a glacially breached watershed named Lairig Gartain.

Paraglacial processes

Earth-surface processes that occur as a result of the rapid readjustment of glaciated landscapes to non-glacial conditions. Distinctive **sediment–landform associations** occur on hill slopes and valley floors, particularly as a result of reworking of unconsolidated, unstable sediments by fluvial activity (Figure P.5).

FIGURE P.5   A paraglacial landscape at Hayeswater, eastern Lake District, England. A plateau glacier of the Younger Dryas age on the fells to the right (High Street range) descended into the lake basin depositing a complex suite of recessional moraines. The gully may represent an ice marginal meltwater channel, hence producing the large alluvial fan below it. Substantial gullying of moraine and debris flows have affected the former subglacial terrain on the slope to the right of the gully.

Parallel roads (a Scottish Highlands term; see **glacial-lake shorelines**)

Parasitic folds

Small folds on the limbs of a larger fold. In glaciers, parasitic folds are typically associated with **similar folds** of several or tens of metres amplitude but an order of magnitude smaller (Figure P.6).

FIGURE P.6   Parasitic folds near the hinge zone of a larger fold of several metres amplitude, Midtre Lovénbreen, northwestern Spitsbergen, Svalbard (August 2013).

Paternoster lakes (see **chain lakes**)

Penitentes (derived from the Spanish)

Groups of spike-like irregularities on the surface of a **glacier**. They may reach heights of several metres, and are taller than they are wide. They are found where intense solar radiation at sub-zero temperature favours sublimation and, consequently the amplification of surface irregularities (Figure P.7). Penitentes represent a more extreme form of differential ablation than **ice ships**.

FIGURE P.7 A group of penitentes on a small glacier above Fossil Bluff Station (UK), Alexander Island, southwestern Antarctic Peninsula (December 2012).

Percolation zone

That part of a **glacier** where meltwater or rainwater percolates into snow or firn. This water may refreeze, forming **superimposed ice** (see also **facies (snow)**).

Periglacial processes

The processes of freezing and thawing of the ground that result in substantial modification of the surface through movement of grains and clasts in soil and unconsolidated sediment, commonly in the presence of water. A variety of landforms are produced including solifluction lobes, **felsenmeer**, stone stripes, stone polygons and turf hummocks, which are not restricted to glacial environments (Figure P.8). In fact, it is estimated that periglacial processes affect a third of the Earth's land surface. The term 'periglacial' was formerly restricted to processes operating at the periphery of former ice sheets (*cf.* **paraglacial processes**; **permafrost**).

FIGURE P.8 Periglacial processes involving solifluction and stone-sorting on a gently sloping surface near Stagnation Glacier, Bylot Island, Canadian Arctic. This surface has a well-established cover of vegetation, on ground that has been ice-free for several tens of thousands of years. Thawing of the underlying permafrost allows slow movement of the water-saturated active layer in summer.

Permafrost

Ground which remains permanently frozen. It may be hundreds of metres thick with only the top few metres or less thawing out in summer (Figure P.9).

FIGURE P.9 An example of permafrost terrain, with typical patterned ground resulting from freeze–thaw processes, Blomstrandhalvøya, Kongsfjorden, northwestern Spitsbergen, Svalbard.

Piedmont glacier (*cf.* ice piedmont)

A **valley glacier** that expands into a lobe-shaped lowland terminal zone after leaving the confines of its valley (Figure P.10a and b).

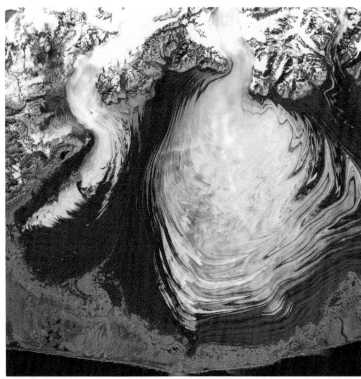

(a)

FIGURE P.10 Examples of large and small piedmont glaciers. (a) Landsat 7 satellite image of the 65 km wide temperate Malaspina Glacier, southeastern Alaska. Vegetation is shown in red, and bare glacier ice (with folded medial moraines) in blue. (Image from NASA, 31 August 2000.) *(Continued)*

P

(b)

FIGURE P.10 (Continued)   (b) Several polythermal piedmont glaciers, some coalescing, in southern Axel Heiberg Island, Canadian Arctic (August 1977).

Piedmont moraine

Distinctive, bulbous ice-cored moraines represented the outer limit reached by piedmont glaciers (Figure P.11).

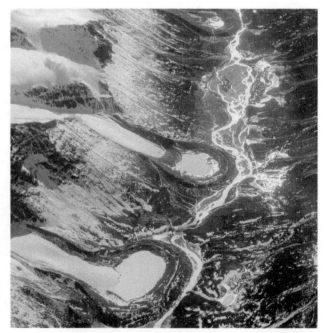

FIGURE P.11   A series of four piedmont moraines formed by small mountain glaciers extending to the floor of a larger glacial trough, northwestern Baffin Island, Canadian Arctic. Each moraine encloses a small lake, the surface of which is frozen (July 2014).

Pinning point

A place where the terminus of a **tidewater glacier** or **ice shelf** remains stationary for long periods during an advance or retreat. In the case of tidewater glaciers, this occurs where a glacial trough shallows, narrows or bends. Ice shelves are pinned by islands or promontories. In such cases, the terminus does not respond directly to **mass-balance** changes. An increase in mass results in increased **calving** from the same position. A negative mass balance results in glacier thinning over a deeper basin behind the pinning point until the whole glacier tongue is destabilised, and the glacier then retreats catastrophically to its next pinning point. Glaciers with pinning points therefore change in response to topographic geometry, rather than directly to climatic change.

Pitted outwash (see **kame**)

Plasticity

The relationship between strain rate and shear stress, whereby the application of stress to a material results in no change until a critical level called the yield stress is reached. At this point, the material undergoes irrecoverable strain, at whatever rate is necessary to prevent the applied stress exceeding the yield stress. At stresses typically operating in glaciers, ice behaves as an approximately plastic material (Figure P.12).

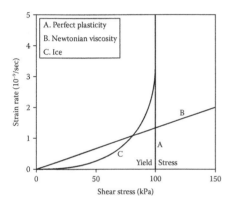

FIGURE P.12   The rheological behaviour of different materials. The line A represents a perfectly plastic material with a yield stress of 100 kilopascals. Line B represents the relationship for a Newtonian linear-viscous material, for which the strain rate is linearly proportional to the shear stress. The curve C represents a non-linear viscous material, such as ice. (Modified from Paterson, W.S.B., 1994. *The Physics of Glaciers*. Pergamon, New York.)

P

Plateau ice cap or plateau icefield

A body of glacier ice that forms on a high-level land mass of subdued or flat topography (Figure P.13). Ice descends the bounding steep slopes and cliffs to forms **cirque** and **valley glaciers** or avalanche cones.

FIGURE P.13   A small plateau ice cap on a promontory near the northern end of George VI Ice Shelf, Antarctic Peninsula. Ice from this plateau feeds the low-level cirque glacier below. The ice shelf and Alexander Island are in the background (November 2012).

Pleistocene

The geological epoch that is dominated by glacial–interglacial cycles. It is the first Epoch of the **Quaternary** Period, the second being the **Holocene**. As defined by the International Stratigraphic Commission, the Pleistocene and Quaternary begin at 2.588 million years, and the Pleistocene ends at 12,000 years BP (Figure P.14).

Era	Period	Epoch	Age (Ma)
Cenozoic	Quaternary	Holocene	
			0.012
		Pleistocene L M E	0.126
			0.781
			2.588
	Neogene	Pliocene	5.332
		Miocene	23.03
	Palaeogene	Oligocene	33.9
		Eocene	55.8
		Palaeocene	65.5

FIGURE P.14   The Cenozoic timescale, showing the context of the Pleistocene Epoch and Quaternary Period within it, as formally defined by the International Stratigraphic Commission in 2009.

Plucking (see **quarrying**)

Polar glacier (see **cold glacier**)

Polish (see **abrasion, Figure A.3**)

Polynya (derived from the Russian)

An area of open water surrounded by continuous **pack ice** (Figure P.15), dotted with loose ice **floes**. Polynyas range in size from a few hundred metres to many kilometres. They are formed either by (a) upwelling of warm water which keeps the surface water above the freezing point, or (b) **katabatic winds** or ocean currents that act to drive newly formed ice away from a fixed location, such as away from the coastline or from an **ice shelf**.

FIGURE   P.15 North Water Polynya, located between Ellesmere Island (illustrated here) and Greenland in northern Baffin Bay. It is the largest polynya in the Arctic, covering about 85,000 km², and is bordered by sea ice adhering to coastal glaciers. It is a biologically very productive region, providing sustenance to algae, fish and marine mammals (June 1975).

Polythermal glacier

A **glacier** that contains both cold ice (below the pressure melting point) and temperate ice (at the pressure melting point) (Figure P.16). The proportion of the different types of ice varies considerably, but typically temperate ice is found at the base of a glacier (where it is warmed by geothermal heat), with cold ice above, and at the snout and margins. Large parts of **ice sheets** are also polythermal in the broad sense, since their base is also at the pressure melting point.

FIGURE   P.16 Two typical polythermal glaciers in northwestern Spitsbergen, Svalbard: Austre Lovénbreen (left) and Midtre Lovénbreen (right). The margins and termini of these glaciers are frozen to the bed. As they become thinner, the proportion of temperate ice decreases.

Portal (see **glacier portal**)

Pothole

A deep cavity, typically cylindrical in shape, formed by the action of debris-laden water under high pressure, such as at the base of a temperate glacier. Rounded stones, which aided the erosion, commonly remain in the base of the hole (Figure P.17).

P

FIGURE P.17 One of several large potholes formed in gneiss bedrock at the end of the last Pleistocene glaciation at Cavaglia, Switzerland. The group of potholes is located on a prominent riegel at an altitude of roughly 1700 m.a.s.l.

Precession of the equinoxes

The wobble of the Earth's axis of spin because of the gravitational pull exerted by the sun and moon. This results in changes in the rotation of the spin axis as the planet orbits the sun on a 19,000 to 23,000 year cycle. The consequence of these changes is that the seasons (or the equinoxes) move around the sun on a regular basis. This is one of three astronomical variables that affect the amount of solar radiation received at the Earth's surface, and forms part of the Astronomical Theory of Climate Change.

Precipitate (see calcium carbonate precipitation)

Pressure melting point

The temperature at which ice melts under a specific pressure. Pressure lowers the melting point to below 0°C beneath a glacier.

Pressure release

The fracture of bedrock along joints following exposure after being covered by glacier ice. Following erosion of valley side by **glaciers**, joints form parallel and normal to the slope, and these fail when the restraining ice is removed. Granite is particularly susceptible to this form of erosion (Figure P.18).

FIGURE P.18   Etive Slabs, above Loch Etive in the Grampian Highlands of Scotland. Joints are parallel and normal to the valley side in this exposure of Caledonian granite.

Pressure ridge

A ridge or wall of broken floating ice forced upwards by wind-induced pressure. Commonly reaching heights of several metres and stretching for hundreds of metres, pressure ridges involve sea ice alone (Figure P.19a), glacier ice and sea ice in combination (Figure P.19b), and glacier and lake ice together (Figure P.19c).

(a)

(b)

FIGURE P.19 Different types of pressure ridge. (a) Sea ice pressure ridge in southeastern Weddell Sea with RV *Polarstern* in the background (January 1991). (b) Sheets of lake ice uplifted where George VI Ice Shelf impinges on permanently frozen Ablation Lake, Alexander Island, Antarctica (November 2012).   *(Continued)*

(c)

FIGURE P.19 (Continued)   (c) Sea ice thrust up against the margin of McMurdo Ice Shelf, near Scott Base (NZ), Antarctica, with Mt Discovery in the background (December 2010).

Primary structures

Structures in glaciers that are the result of accumulation and accretion, in contrast to **secondary structures**, which are the product of deformation. The principal primary structure is **stratification**, as well as **ice lenses**, regelation layering and breccia; each of these is described and illustrated separately.

Proglacial area (see **glacier forefield**)

Proglacial channel

The river channel that begins at a glacier **snout** and flows out over the ground in front. Within a **temperate glacier**, a well-developed **subglacial stream** emerges as a single channel (Figure P.20) and may or may not become braided downstream. Channels are lined with river-transported cobbles and boulders, and migrate across the proglacial zone unless strongly constrained by topography.

FIGURE P.20   Single channel associated with water discharge from temperate Franz Josef Glacier, South Island, New Zealand (January 2011).

Proglacial lake

A lake developed immediately in front of the glacier, commonly bordered by the mounds of unconsolidated deposits that characterise the terminal zone of a glacier (Figure P.21).

(a)

(b)

P

FIGURE P.21 Aerial views of contrasting proglacial lakes. (a) Bear Lake on the Kenai Peninsula of Alaska, enclosed by a prominent terminal moraine, and carrying numerous icebergs (July 2011). (b) Rock basin-constrained lake in front of Fingerbreen, an outlet glacier from the western ice cap of Svartisen, northern Norway (August 1971).

Pure shear

One of the two types of **strain**, pure shear involves flattening or stretching of a material under compressive or tensile (extending) stress (*cf.* **simple shear**). If represented by a strain ellipse (as is common in structural geology), the lines drawn normal to the principal stresses retain the same orientation throughout the deformation (Figure P.22). Pure shear is also referred to as coaxial strain.

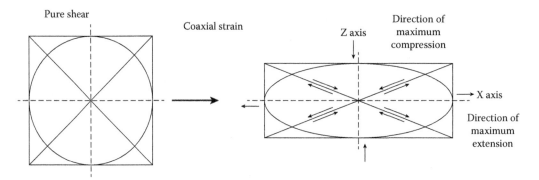

FIGURE P.22 Diagram to illustrate the effect of pure shear on a square and circular object, in relation to the principal stress axes.

Push moraine

A ridge of debris pushed up by a **glacier** as it advances into its **proglacial zone**. The ridge is heterogeneous, with a mix of basal and supraglacial debris, and fluvial, lacustrine or marine sediment. Temperate glaciers bulldoze the sediment in front (Figure P.23a), whereas polythermal glaciers in addition deform the frozen ground (permafrost) in front, producing multiple ridges (Figure P.23b). Thrust moraines are a type of push moraine.

(a)

FIGURE P.23 Push moraines associated with a temperate and a polythermal glacier. (a) A push moraine, a few metres high, at the snout of temperate Brikdalsbreen during its most recent advance. Brikdalsbreen is an outlet glacier from the Josterdalsbreen ice cap in central southern Norway (June 1996).   *(Continued)*

(b)

FIGURE P.23 (Continued)   (b) A massive push moraine in front of Thompson Glacier, Axel Heiberg, Canadian Arctic. This polythermal glacier is associated with permafrost, and stress imposed on this frozen sediment results in brittle failure and a series of stacked ridges (July 2008).

Pyramidal peak

A steep-sided peak that is approximately pyramidal in shape, which has been subject to glacial erosion on three or four sides (Figure P.24). The steepest pyramidal peaks are referred to as **horns**.

P

FIGURE P.24   The pyramidal peak of Nevado Santa Cruz, Cordillera Blanca, Peruvian Andes (January 2002).

Pyroclastic deposits

The products of explosive volcanic eruptions, falling out from the eruption column and from fast-flowing dense clouds of ash, or as debris flowing downslope especially when mixed with water and melting ice. Pyroclastic deposits may cover a glacier located on a volcano or within its crater (Figure P.25).

FIGURE P.25   Pyroclastic deposits on a small glacier within the summit crater of Ruapehu, North Island, New Zealand. The grey ash dates from the eruption of 2007, and the photograph was taken in April 2008.

P

Q

Quarrying

The process of bedrock fracture and removal beneath a sliding glacier. Transitory stress concentrations produced, for example, by embedded stones within the ice, may enlarge cracks and displace fragments that are then incorporated in the base of the glacier (Figure Q.1). The process may be facilitated by prior freeze–thaw bedrock fracturing, aided by high basal water pressures.

FIGURE Q.1 Bedrock fracture, displacement and transportation of clasts in a cave beneath Grosser Aletschgletscher, Valais, Switzerland (October 2010).

Quaternary Period

The last period of geological time, representing the last 2.588 million years (see **Pleistocene** for discussion of timescale).

Quiescent phase

The period in which a **surge-type glacier** is slow-moving or stagnant (Figure Q.2). This period is typically decades long in contrast to the surge phase that may last only a few months or years.

FIGURE Q.2   Steele Glacier is a 40 km long surge-type glacier in the Icefield Ranges, Yukon. It last surged during 1965–1966, leaving a long tongue of debris-mantled stagnant ice during the current quiescent stage (July 2006).

Q

R

Raft (also known as a **megablock**)

A large block of bedrock detached from the substrate beneath a **glacier**, occurring within or adjacent to associated **glacigenic** or other sediments (Figure R.1). A raft may be transported by ice for a considerable distance (tens or hundreds of kilometres) and is emplaced by glaciotectonic processes, yet retains its original lithological appearance.

FIGURE R.1 Large raft of locally derived Cretaceous chalk at Overstrand, Norfolk, England. It has been raised several metres above the host bedrock and emplaced by glaciotectonic processes, along with sand and gravel. The base of the raft is defined by a thrust. The deposit was emplaced during the Middle Pleistocene Anglian glaciation. (Photograph by Emrys Phillips.)

Raised beach

A beach that has been uplifted as a result of the land responding to ice-unloading (**isostatic rebound**) following the disappearance of an **ice sheet** (Figure R.2), or lowering of sea level as an ice sheet grew in another location. Well-developed raised beaches are commonly backed with cliffs.

FIGURE R.2 Raised beach at Gruinard Bay, Northwest Highlands, Scotland. It was formed when relative sea level was higher after the end of the last glaciation. The beach was raised as a result of isostatic rebound after the ice sheet over Scotland disappeared.

Randkluft (derived from the German)

A deep, narrow cleft that develops between a rock face and steep **firn** and **glacier ice** at the head of a glacier (Figure R.3). It is formed as a result of snow and ice melting against warmer rock. Freeze–thaw processes are important in such locations, and with quarrying and removal by ice are effective zones of glacial erosion. The term is sometimes used more loosely for the gap between the sides of a glacier and the bedrock, but this usage is not favoured here (*cf.* **bergschrund**).

FIGURE R.3 Climbers crossing the Randkluft at the top end of Höllentalferner. Although only 0.223 km² in area (2010), it is the largest glacier in Germany, located on the north flank of the Zugspitze, at 2962 m on Germany's highest mountain. (Photograph by Uli Herzog, September 2008.)

Rat-tail

A minor ridge in bedrock, parallel to striations, extending down-glacier from a knob of more resistant rock (Figure R.4).

FIGURE R.4 Rat-tail on a striated basalt surface, formed down-glacier of a harder quartz crystal, Skálafellsjökull, southeastern Iceland.

Raymond bump (named after an American glaciologist)

An up-warped arch of annual accumulation layers in an **ice sheet** or **ice rise**, formed at an **ice divide** in response to vertical changes in **strain** (Figure R.5). Raymond bumps have been identified by ice-penetrating radar. They may form symmetrically in association with a slight elevation change in the bedrock beneath, but their morphology may change if the ice divide migrates in response to changing ice-sheet dynamics.

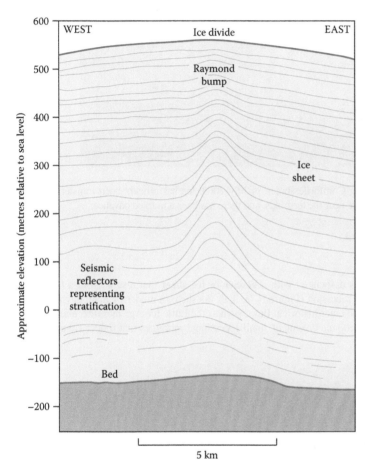

FIGURE R.5 Interpretation of a radargram of a Raymond bump in internal ice-layering on Roosevelt Island, an ice rise in the Ross Sea, Antarctica. (Redrawn and simplified from an original radar image in: Martin, C., et al., *JGR Earth Surface*, 111, 2006, DOI: 10.1029/2005JF000326.)

Reaction time

The time required for the **snout** or terminus of a glacier to respond to changes in **mass balance**. A change to a positive mass balance may result in a glacier advance on a decadal timescale for **alpine glaciers**, or a centennial or longer timescale for **ice sheets**. Gradient also plays a part: long, low-gradient glacier tongues may not respond to an increase in mass unless sustained over several decades, whereas short steep glaciers may respond with an advance within a few years.

Receiving area

The lower reaches of a **surging glacier** that receives an increase in mass through transfer from the **reservoir area** (Figure R.6). During the **surge**, ice in the receiving area thickens considerably, which may result in a dramatic advance of the terminus, and at the same time the reservoir area is subject to surface lowering. The transfer can be as rapid as a few months for **temperate glaciers** (as in Alaska and the Yukon), or take a few years for **polythermal glaciers** (as in Svalbard).

FIGURE R.6   Aerial view of surging Comfortlessbreen in July 2009. Ice has already been transferred to the receiving area, the heavily crevassed ice near the terminus, and there has been a marked advance.

Recession (see **glacier recession**)

Recessional moraine

A glacially deposited ridge of debris associated with a **glacier** or **ice sheet** that results from a temporary readvance during a period of overall recession (Figure R.7a, b, c). As such, in modern glacial environments, a recessional moraine lies between the outer historical limit, such as the Little Ice Age moraine, and the modern glacier margin. Ice sheets may produce dozens of small recessional moraines that reflect changes in ice volume over many millennia.

(a)

FIGURE R.7   A range of modern and ancient recessional moraines. (a) The proglacial area of Gaißbergferner (also known as Gaisbergferner), Ötztaler Alpen, Austria. The lower moraine is the terminal moraine of the Little Ice Age, and the mid-valley moraine is the mid-20th century recessional moraine (September 2014).

(Continued)

(b)

(c)

FIGURE R.7 (Continued) (b) A series of closely spaced recessional moraines, comprising angular debris transported by cold ice, associated with a tongue of ice from the East Antarctic Ice Sheet, encroaching Roberts Massif, at latitude 85°S. (December 1995). (c) A series of recessional moraines dating from the Younger Dryas (c. 12,000 BP), associated with a plateau icefield feeding a valley glacier at Hayeswater, Lake District, England.

Reconstructed glacier (see **regenerated glacier**)

Recrystallisation

The transformation of grains of snow, **firn** or old ice crystals to a new assemblage of ice crystals, without any melting (Figure R.8). The process results in changes to mean crystal size and orientation (see **crystal fabric**), and an increase in density (see also **ice crystal**).

FIGURE R.8   Schematic diagram illustrating how snow is transformed first into firn, and then glacier ice by recrystallisation.

Recumbent fold

A fold that has an axis or axial plane that is almost horizontal, that is, the fold is lying on its side, and is the product of drag of basal layers over the bed. Typically, recumbent folds occur in the basal layers of a **glacier** or **ice sheet** and also involve **basal debris** (Figure R.9a and b). See **fold** for explanation of terminology.

(a)

FIGURE R.9 (a) Large recumbent fold in a polythermal glacier, comprising foliation and basal ice layers in a longitudinal cliff section at Crusoe Glacier, Axel Heiberg Island, Canadian Arctic (July 2008). (b) Recumbent fold involving basal debris in the base of Meares Glacier, a temperate tidewater glacier in southern Alaska (June 2012).

(b)

Reference glacier (see **index glacier**)

Refolded fold

Reflecting multiple phases of deformation, a refolded fold is a product of two phases of folding, recognised because the axial plane of the first fold has itself been folded (Figure R.10). In glacier ice it is unusual to observe refolded folds because of recrystallisation.

FIGURE R.10 Refolded fold at the confluence of Thompson Glacier and White Glacier, Axel Heiberg Island, Canadian Arctic (Summer 1975). Note how the black debris layer, forming an isoclinal fold with two parallel limbs with its apex (hinge) to the right is folded itself by the more dominant fold.

Regelation

The process whereby ice is formed from meltwater as a result of lowering of pressure beneath a **glacier** (Figure R.11). It is commonly accompanied by **pressure melting** such as on the up-glacier side of a bump; the water then flows around the bump and refreezes. This ice is referred to as **regelation ice**.

FIGURE R.11 Regelation ice in the basal debris layer of Fountain Glacier, Bylot Island, Canadian Arctic (July 2014).

Regenerated glacier (or rejuvenated glacier/reconstructed glacier)

A glacier that develops from ice-avalanche debris beneath a rock cliff (Figure R.12).

FIGURE R.12   Suphellebreen, a regenerated glacier below an outlet glacier from the Josterdalsbreen ice cap, southern Norway. The glacier was advancing during this July 1996 photograph, when a party of glaciologists was inspecting the glacier.

R

Remnant glacier

A glacier that still survives, even though it no longer has an **accumulation area**, a situation that arises when the **equilibrium line** has risen to a level above the glacier as a consequence of climatic warming (Figure R.13).

FIGURE R.13   A large and a small remnant glacier above an ice-dammed lake between Aktineq Glacier (background) and Fountain Glacier (foreground), Bylot Island, Canadian Arctic. Note the absence of any accumulation area on these two glaciers (July 2014).

Reservoir area

The upper reaches of a surge-type glacier where ice slowly builds up without being transferred down valley until a surge takes place (Figure R.14).

FIGURE R.14   The reservoir area of upper Comfortlessbreen, northwestern Spitsbergen, Svalbard, during a surge, when ice is in the process of being transferred to the glacier tongue (the receiving area). Note the scar beneath the two pointed peaks and ridge in the background, which separates static ice from active ice and indicates surface lowering in progress (July 2009).

R

Response time

The time required for a change in **mass balance** of a glacier to be manifested in change in the geometry of the glacier, particularly the position of the **terminus**. The reaction time of a glacier is dependent on the extent to which it is out of equilibrium and requires continuing observation over decades.

Retreat (see **glacier recession**)

Ribbed moraine (see **Rogen moraine**)

Ribbon lake

A long, narrow lake resulting from glacial erosion, and commonly containing multiple basins (Figure R.15).

FIGURE R.15 The ribbon lake of Ullswater in the English Lake District with its notable 'dog-leg' reflecting a change in ice-flow direction, viewed from Gowbarrow Fell.

Riegel (derived from the German)

A rock barrier that extends across a glaciated valley, usually comprising harder rock, and often having a smooth up-valley facing slope and a rough down-valley facing slope (Figure R.16).

FIGURE R.16 The glacier-fed lake in front of Triftgletscher, Switzerland, with a prominent riegel in granite beyond (August 2007).

Rift

A large fracture that occurs in floating masses of glacier ice such as **ice shelves** and **glacier tongues**. Rifts occur particularly towards their seaward limit, and ultimately produce large **tabular icebergs** (Figure R.17a and b).

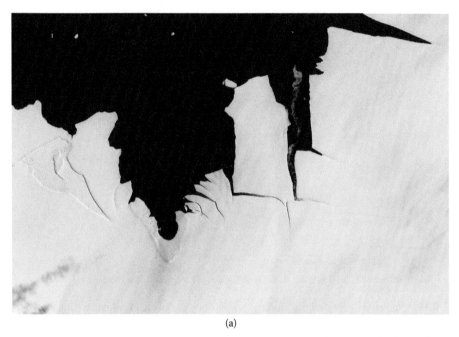

(a)

(b)

FIGURE R.17 Satellite and ground images of rifts at the edge of King Baudouin Ice Shelf, East Antarctica. (a) Landsat 8 image dated 24 January 2015 showing extensive rift development and the birth of a 70 km² tabular iceberg. The image measures about 40 km across. (Image from NASA kingbaudouin_oli_2015024_lrg.) (b) Travelling across sea ice between the walls of a rift at the western side of the large bay shown in the satellite image. (Photograph by Bryn Hubbard, 2010.)

Rill

The small channel produced by running water on the surface of a glacier or sediment, usually a few centimetres across and ephemeral (Figure R.18).

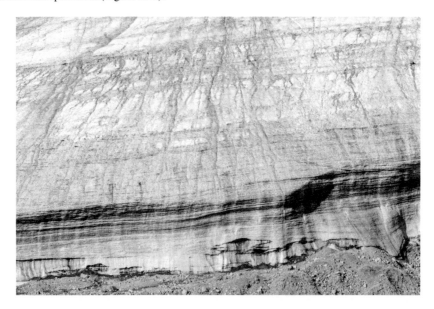

FIGURE R.18   Numerous rills draining surface water from the snout of Fountain Glacier, Bylot Island, Canadian Arctic. Some rills join large supraglacial streams, but many flow parallel to fractures and over the edge of the marginal cliff (July 2014).

Rime

Ice with trapped air, formed on an exposed surface as a result of rapid freezing of very small supercooled water droplets in the atmosphere (Figure R.19) (*cf.* **hoar frost**).

FIGURE R.19   Rime on the supporting structures of a meteorological station near Luosto, northern Finland.

Rinnentaler (derived from the German; see **tunnel valley**)

Roche moutonnée (from French; plur. **roches moutonnées**)

A rocky hillock with a gently inclined, smooth up-valley facing slope resulting from glacial abrasion, and a steep, rough down-valley facing slope resulting from quarrying by the glacier (Figure R.20). A **flyggberg** is a larger version of this feature.

FIGURE R.20   A group of roches moutonnées carved out of granite at the newly formed proglacial lake of Rhonegletscher, Valais, Switzerland (September 2009).

Rock avalanche

A sudden rockfall event of large magnitude, commonly resulting from slope destabilisation by erosion or an earthquake in a tectonically active region (Figure R.21). Rock avalanches are classified as 'low frequency, high magnitude' events. If a rock avalanche covers part of a glacier, it can retard ablation and may temporarily lead to a glacier advance.

FIGURE R.21   Rock avalanche deposits on Buckskin Glacier, Alaska Range. The marked step at the edge of the deposit is caused by faster ablation of the bare ice around it (July 2011).

R

Rock basin

A lake- or sea-filled bedrock depression carved out by a **glacier** (Figure R.22).

FIGURE R.22 Two rock basins, last cut in gabbro by glaciers of the Last Glacial Maximum and Younger Dryas, Cuillin Hills, Isle of Skye, Scotland. The large basin is Loch Coruisk, a lake whose surface is a few metres above sea level, and the basin to the left is a marine inlet. The two are connected by a short river incised in a bedrock riegel.

Rock drumlin (see **whaleback**)

Rockfall/rockslide (process and deposit)

Rockfalls are 'high frequency, low magnitude' events in contrast to **rock avalanches**. In glacial environments, freeze–thaw processes trigger many rockfalls, leading to accumulations of **supraglacial debris** at the margins of a **glacier** (Figure R.23). Rockslides are similar to **rock avalanches** in being 'low frequency, high magnitude' events but are specifically the result of failure along a plane of weakness in the rock mass, such as a bedding plane or joint. Occasionally, during the fall, debris becomes airborne and accumulates as a pulverised mass, far distant from the source.

FIGURE R.23 Rockfall below the peak of Piz Bernina, adding supraglacial debris to the surface of Vadret da Morteratsch, Switzerland (September 2009).

Rock flour

Bedrock that has been pulverised at the bed of a **glacier** into clay- and silt-sized particles. It is commonly carried in suspension in glacial meltwater streams, which consequently take on a milky appearance (*cf.* **glacier milk**) (Figure R.24).

FIGURE R.24 Rock flour being transported by the outflow from Hooker Glacier with Aoraki/Mt Cook in the background, South Island, New Zealand (January 2011).

Rock glacier

A mass of rock debris with interstitial ice (ice between boulders), showing evidence of slow **creep**, such as arcuate surface ridges of debris (Figure R.25a and b). The term is sometimes used more loosely for **glaciers** that are debris-mantled from head to toe.

(a)

FIGURE R.25 (a) Rock glacier in Val Muragl, Switzerland, with arcuate ridges indicating slow creep of the rock-ice mass. Velocities up to half a metre per year have been measured (September 2011).

(b)

FIGURE R.25 (b) Rock glacier descending from Sourdough Peak, Wrangell Mountains, Alaska. It is more than 3 km long and has a maximum width of nearly 1 km (July 2011).

Rockslide (see **rockfall**)

Rock-slope failure

The detachment of a large slab or slabs of rock from intact bedrock followed by downslope movement or collapse. A continuum exists from catastrophic failures (including **rock avalanches**), through arrested rockslides, to almost *in situ* slope deformation. The 'oversteepened' sides of glacial valleys are particularly prone to rock-slope failure (Figure R.26a), and these may occur up to several millennia after deglaciation (Figure R.26b).

(a)   (b)

FIGURE R.26 Contrasting types of rock-slope failure. (a) Recent failure within a limestone cliff above the rapidly receding, debris-mantled snout of Unterer Grindelwaldgletscher, Switzerland (October 2006). (b) Context of a late glacial rock-slope failure where two glacial valleys meet in the northern Lake District, England. The rounded bulbous shape below the shadowed cliffs of Clough Head is the failure, and it is characterised by transverse fractures with depressions. The photograph was taken from the nearby mountain of Blencathra, looking over the village of Threlkeld.

Rogen moraine (named after a Swedish lake; also known as Ribbed moraine in North America)

A landscape of numerous irregular ridges of sediment, aligned roughly transverse to glacier flow (Figure R.27). Ridges are typically up to 30 m high and 100 m long. Some ridges are broadly arcuate in shape with their 'horns' pointing down-glacier. Individual ridges are asymmetric in cross-section, with a gentle slope facing up-glacier and a steep face down-glacier. Lakes commonly surround the ridges. Sediments in the ridges are variable, and include diamicton, sand and gravel, commonly showing signs of deformation. Rogen moraines are formed subglacially, but are of complex origin, involving folding, thrusting and stacking of basal or subglacial sediment.

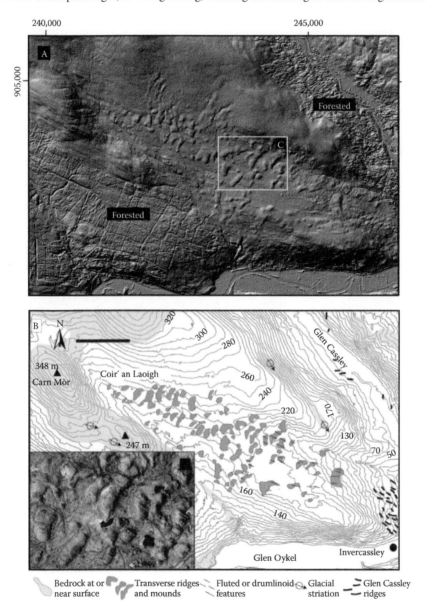

FIGURE R.27 Characteristics of Rogen moraine in Coir' an Laoigh, Scotland. (A) Hill-shaded digital surface model, constructed from Intermap Technologies' NEXTMap Britain topographic data, showing transverse ridges with illumination from the northwest. (B) Geomorphological map of the Rogen moraine area, with contours at 10 m intervals. (C) Aerial photograph showing development of superimposed streamlined landforms in the southeast (bottom right) (figure from Andrew Finlayson, British Geological Survey).

Röthlisberger channel (named after a Swiss glaciologist; also known as **R-channel**)

A channel at the base of a glacier that is incised upwards into the ice (Figure R.28). The channel may be one of several forming and anastomosing network beneath a glacier, typically one that is temperate (*cf.* a **Nye channel**, which is incised into bedrock).

FIGURE R.28   A large Röthlisberger channel, with semi-circular cross-section, incised into the base of Pastoruri, an ice cap in the Cordillera Blanca, Peru (2002).

Runoff (see **discharge**)

(a) The flux of water leaving a glacier. (b) The **discharge** of water from a drainage basin (with or without a glacier) at a measured cross-section, divided by the area of that basin, expressed as millimetres of water equivalent per day, or kilograms per metre-squared per second.

Run-out distance

The distance an ice avalanche travels from its source (Figure R.29a and b). Run-out distances of ice avalanches are difficult to predict, as they depend not only on the volume of ice descending but also on the shape and roughness of the terrain in the avalanche path. When travelling over a homogeneous, smooth firn surface (as in the example shown in Figure R.29a and b), the horizontal distance the ice avalanches travel may be up to three times longer than the altitude difference between the highest point of the starting zone and the lowest point of the deposit.

(a)

(b)

FIGURE R.29   A large ice avalanche from a hanging glacier on the south flank of Mönch, Berner Oberland, Switzerland. (a) Aerial photograph taken on 6 July 1984, a day after the avalanche. The run-out distance reached almost 700 m. (b) The frontal part of the avalanche as seen from the ground. The run-out distance was so long that parts of the avalanche covered a path frequently traversed by tourists visiting an Alpine Club hut near Jungfraujoch. Fortunately, no one was injured in this event.

S

Sand

Sedimentary particles ranging from 62.5 μm to 2 mm (or 4 to −1 on the φ scale). Sand in glacial environments is the product of a wide range of processes. Bedrock erosion by glaciers, especially in areas of granite and gneiss, produces a copious amount of sand, which is incorporated in basal glacial deposits (**till**) and reworked by **glaciofluvial** activity. Well-sorted sand occurs in patches on the bed of glacial streams (Figure S.1).

FIGURE S.1   Sand deposited during a phase of slack flow by the stream emanating from Vadrec del Forno, Switzerland. The glacier is visible in the background (September 2011).

Sandur (plur. Sandar) (from the Icelandic)

Extensive flat plain of sand and gravel with **braided streams** of glacial meltwater flowing across it (Figure S.2). Sandar are usually not bounded by valley walls and commonly form in coastal areas.

FIGURE S.2   Sandur in front of the large outlet glacier, Skeiðarárjökull, southern Iceland (March 2015).

Sastrugi (derived from transliterated Russian)

Wind-carved snow forming irregular ridges, resembling sand dunes. They typically have a relief of tens of centimetres to a metre or more, and are commonly aligned in the direction of the prevailing wind (Figure S.3). Sastrugi is typical of the accumulation areas of **ice caps** and **ice sheets**, and also occurs off-ice in the surrounding terrain and other snow-covered areas.

FIGURE S.3 Sastrugi on the upper edge of a cirque on Fisher Massif, overlooking the 80-km wide the Amery Ice Shelf–Lambert Glacier system, East Antarctica. The prevailing wind is from right to left (December 1994).

Scablands (see **channelled scablands**)

Scalloped ice

(a) An assemblage of small hollows in ice, resulting from uneven ablation during intense solar radiation. They are typically 5–20 cm in diameter and a few centimetres deep (Figure S.4a). (b) A similar formation, but the result of strong winds and eddying in the lee of surface irregularities. On a glacier surface they are of similar size to those from solar radiation, but in **ice caves** may be a metre or more in diameter (Figure S.4b).

(a)   (b)

FIGURE S.4 (a) Scalloped ice at the surface of the temperate Franz Josef Glacier, South Island, New Zealand (April 2008). (b) Metre-diameter scallops formed by wind eddying in an ice cave beneath Breiðamerkurjökull, southern Iceland (March 2015).

Scoop

An unusual glacial erosional feature, shaped as a partially developed basin a few tens of centimetres wide, carved out of bedrock at the base of a sliding glacier (Figure S.5). A steeper, sharp-edged face occurs at the up-glacier end, and the feature itself may be striated.

FIGURE S.5   Scoop and associated striations in basalt bedrock at Skálafellsjökull, southeastern Iceland. Ice flow was from right to left. Coin for scale.

Scree (see also **talus cone**)

An accumulation of loose stones on a steep mountainside. Scree is the product of freeze–thaw processes in cliffs above, and the detached blocks fall, roll or slide downslope (Figure S.6a and b). Scree consists mainly of boulders and cobbles, typically resting at an angle of 32°–35°. Where actively forming, scree slopes are unstable when walking on them. Scree formation is an active process in glacial environments, forming fans of debris below gullies, and adds to the supply of supraglacial material on the sides of a glacier. Screes also form along the flanks of glacial troughs following recession of the ice, and in mountainous areas generally.

(a)

FIGURE S.6   Examples of scree slopes in glaciated terrain. (a) Multiple scree cones at the head of Moraine Lake that occupies a large cirque in Banff National Park, Canadian Rockies.   *(Continued)*

S

(b)

FIGURE S.6 (Continued)   (b) The screes that line the glacial trough containing Wastwater, the deepest lake in England.

Sea ice

Ice that forms by the freezing of sea water. The term encompasses a wide range of sea-ice forms, ranging from individual ice **floes** to **fast ice** (Figure S.7) (*cf.* **ice shelves** and **icebergs**, which also float on the sea, but are derived from glacier ice on land).

(a)

FIGURE S.7   (a) Dense sea ice in the southeastern Weddell Sea off the coast of Coats Land, Antarctica. The German icebreaker, RV *Polarstern*, is here temporarily trapped in a mixture of ice floes, pressure ridges and 'porridge ice' (broken floes mixed with new elastic ice and snow) (January 1991).   *(Continued)*

(b)

FIGURE S.7 (Continued)   (b) Solid sea ice ('fast ice') wrapping around the coast of the western Antarctic Peninsula, north of Adelaide Island (October 2012).

Sea loch

The Scottish term for a **fjord**, a sea loch is a glacially eroded trough that typically has an overdeepened basin, with depth exceeding that of the adjacent continental shelf (Figure S.8a). A bedrock sill is commonly found at the mouth of the sea loch, as in fjords generally (Figure S.8b). See also **loch**.

(a)

FIGURE S.8   (a) Aerial view of sea lochs Nevis (upper left) and Hourn (mid-left), along with the deepest lake in Britain, Loch Morar, on the western coast of Scotland. All three lochs are the product of glacial erosion beneath the British-Irish Ice Sheet.   *(Continued)*

S

(b)

FIGURE S.8 (Continued) (b) Converging sea lochs in the Grampian Highlands of Scotland: Loch Leven in the foreground merging with Loch Linnhe beyond. A sill with islands separates the waters of the two lochs.

Secondary structures

A general term to cover all structures in glacier ice that are the result of deformation. They include ductile structures, such as **folds, foliation** and **boudinage**; and brittle structures such as **crevasses, crevasse traces, fractures** and **faults**.

Sediment core

A cylindrical length of sediment extracted from the ground, a lake bed or the sea. Cores are extracted to determine past climate and environmental conditions. Cores may consist of soft sediment for short-term changes (centennial to millennial) or bedrock for long (million year) time scales (Figure S.9). Long sediment cores covering the entire **Cenozoic** glacial history on the Earth complement the several hundred thousand year time scale obtained from **ice cores**.

FIGURE S.9 A half-metre length of sediment core, from the CIROS-1 Drilling Project, McMurdo Sound, Antarctica. The core was split lengthways to log changes in environmental history. The figure on the left (278.32 m) represents the depth below the sea floor in this 702-m long core. It shows a diamictite with a large dolerite clast deposited by glaciers over the site during late Oligocene time (c. 30 million years ago).

Sediment flow (see also **debris flow**)

A general term for movement of saturated sediment down a slope (Figure S.10). Sediment flows commonly comprise sets of lobate forms. Material involved ranges from poorly sorted bouldery sediment (**diamicton**), through **sand** to **mud**.

FIGURE S.10   A sediment flow comprising well-sorted sand, triggered by meltwater flowing off the surface of Wright Lower Glacier, Victoria Land, Antarctica. Pencil for scale lower right (January 1999).

Sediment flushing/evacuation (by water)

The process of removal of large volumes of sediment from a glacier during periods of lake outburst, strong melt or heavy rainfall (Figure S.11).

FIGURE S.11   Sediment flushing from Franz Josef Glacier, South Island, New Zealand, during heavy rain. This area is prone to frequent spells of torrential rain, and this maintains an efficient subglacial drainage system and frequent flushing of sediment (Summer 1987).

S

Sediment–landform association

A characteristic assemblage of landforms and their constituent sediments, associated with a distinctive environment. In glacial environments, such assemblages include glacial, **glaciofluvial**, **glaciolacustrine** and **glaciomarine** depositional features, comprising a wide range of sediment types (**facies**). Different types of glacial sediment–landform associations can be distinguished, according to the climatic and tectonic regimes in which they occur; an example of a polythermal sediment–landform association is shown in Figure S.12 (*cf.* **Landsystem**).

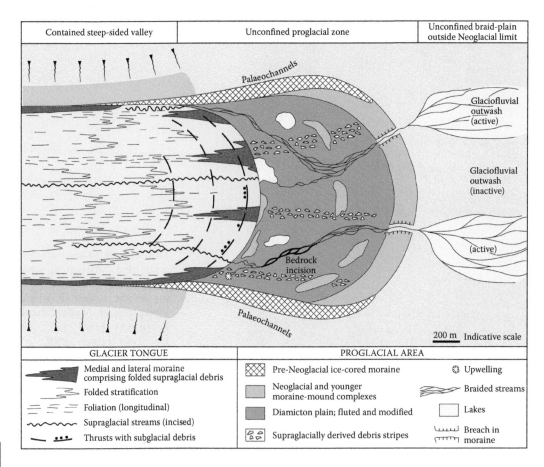

FIGURE S.12   An example of a sediment–landform association typical of a polythermal terrestrial glacier in Svalbard. (From Hambrey, M.J. and Glasser, N.F., *Sedimentary Geology*, 251, 1–33, 2012.)

Sediment plume

An outflow from a glacier of sediment-laden water (Figure S.13). The sediment is carried in suspension into a lake or the sea. Suspended sediment is mainly silt- and sand-sized, and slowly settles on the lake or sea floor. Plumes may occur at the surface of the sea where the sediment-laden water is less dense than saline water, or may hug the sea or lake bottom as **turbidity flows** from which the sediment accumulates as graded beds.

FIGURE S.13 Aerial view of sediment plumes emanating from Frijthovbreen, a tidewater glacier in western Spitsbergen, Svalbard. Different stages of settling-out of sediment are visible in this aerial photograph, the richest orange-brown tones indicating the highest sediment concentration (July 2009).

Sediment ripples

A collection of small ridges of soft sediment (**sand** or **mud**) formed on the bed of a river, lake or the sea. Ripples have wavelengths ranging from a few centimetres to half a metre. Ripples in river beds are asymmetric and their crests are orientated perpendicular to current flow (Figure S.14a). Ripples in lakes or the sea are both symmetric (from wave action) and asymmetric from bottom currents (Figure S.14b). Sediment ripples are characteristic of glacial systems involving flowing water.

(a)

FIGURE S.14 Sediment ripples in glacial environments. (a) Asymmetric current ripples comprising sand with a drape of mud, in the bed of the glacial meltwater stream from Vadret da Morteratsch, Switzerland.   *(Continued)*

S

FIGURE S.14 (Continued) (b) Small symmetric wave ripples of sand representing the exposed bed of a glacier-fed lake that has been lowered for hydro-electric power generation, Oberaargletscher, Switzerland.

(b)

Sedimentary dykes (see **clastic dykes**)

Sedimentary stratification (see **stratification**)

Seepage

The movement of water between coarse particles and eventual emergence at the surface, such as from a glacier through a **moraine** (Figure S.15). Changes in seepage patterns may be indicative of melting of an ice core and point to potential destabilisation of a moraine. This is especially important if the moraine holds back a **proglacial lake**.

FIGURE S.15 Seepage of water from Khumbu Glacier, Everest region, Nepal, through its Little Ice Age moraine, has created a sheet of ice in sub-zero air temperatures (April 2003).

S

Selective linear erosion

A focused form of glacial erosion in valleys beneath an **ice cap** or **ice sheet**. Erosion is concentrated in the valley floors where the ice is thickest and sliding on its bed. The ice is thinnest on the relatively higher ground between valleys, and is frozen to the bed, and therefore this terrain is less susceptible to erosion (Figure S.16).

FIGURE S.16 Selective linear erosion in northeastern Baffin Island, Canadian Arctic. The fjords, covered by sea ice, represent glacial troughs scoured by former ice streams within a large ice cap, while the plateau areas between were little eroded and still retain remnant ice caps, frozen to the bed (July 2014).

Sequence stratigraphy

A branch of geology and geophysics that uses unconformities (breaks in sedimentation) to define and correlate sequences of strata in terms of sediment supply and relative changes in sea level. Sequence stratigraphy can be applied to sequences of glacial sediments, especially in the offshore realm, where a combination of seismic stratigraphy and core stratigraphy can be used to define glacial surfaces of erosion, subglacial and glaciomarine deposition, and changes in sea level (Figure S.17).

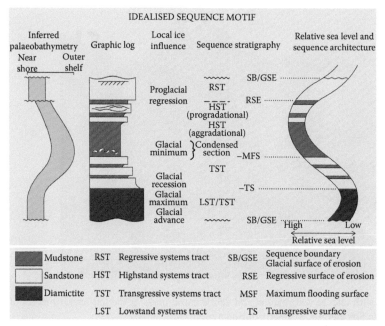

FIGURE S.17 An idealised sequence stratigraphic framework for glacial deposition on a continental shelf where cycles of erosion and deposition by grounded ice are followed by glaciomarine sedimentation and declining glacial influence. This idealised 'sequence motif' is based on seismically constrained drilling by the Cape Roberts Project in the western Ross Sea, Antarctica (From Fielding, C.R., Taish, T.R. and Woolfe, K.J. 2001. Facies architecture of the CRP-3 Drillhole, Victoria Land Basin, Antarctica. *Terra Antartica*, 8, 217–224.).

Sérac (derived from the French)

A tower of unstable ice that forms between **crevasses**, often in **icefalls** or other regions of accelerated glacier flow. Séracs are one of the most dangerous aspects of glaciers for mountaineers owing to their propensity to collapse without warning (Figure S.18).

FIGURE S.18   A zone of crevasses and séracs in an icefall on Fox Glacier, South Island, New Zealand (January 2011).

Shear (see **pure shear** and **simple shear**)

Shear margin

A zone of intense deformation and fracturing at the margin of a **valley glacier**, typically developed during a **surge**, when the velocity gradient between fast-moving ice in mid-glacier and the margin is much greater than normal (Figure S.19).

FIGURE S.19   Shear margin with intense deformation and debris entrainment at the left-lateral margin of Comfortlessbreen, northwestern Spitsbergen, Svalbard, during a surge (July 2009).

S

Shear zone

A zone of severe deformation, especially where a fast-flowing **ice stream** moves past relatively slow-moving ice (Figure S.20). The deformation is characterised especially by a transition from crevasse-free terrain to a zone of intense crevassing. The basal zone of a glacier may also be considered as a shear zone (see **basal ice**).

FIGURE S.20 An unnamed ice stream near the seaward edge of the East Antarctic Ice Sheet, Queen Elizabeth Land, Antarctica. The shear zones on either side of the ice stream are indicated by the transition from crevassed ice to relatively smooth snow-covered ice (November 1994).

Sheath fold

A **fold** that is produced in a shear zone, whereby part of the axial region (fold axis) advances relative to other parts. When examined normal to the fold axis, it has the appearance of an eye (hence also **eyed fold**). In a **glacier**, sheath folds are produced near the base, where simple shear is dominant, and in this situation commonly involves basal debris (Figure S.21).

FIGURE S.21 Sheath fold in the frontal cliff of advancing Crusoe Glacier, Axel Heiberg Island, Canadian Arctic (Summer 1975).

S

Shelf ice

Marine ice, frozen sea spray, glacier ice or ice derived from accumulation of snow that makes up part of an **ice shelf** (Figure S.22).

FIGURE S.22   Shelf ice (frozen sea spray) on the upper surface of Ekström Ice Shelf, near Neumayer Station (Germany) (January 1991).

Sichelwanne (from the German; plur. **Sichelwannen)**

A crescent or sickle-shaped groove formed on bedrock by either glacial erosion or subglacial water under high pressure, or a combination of both. The 'horns' of the crescent shape point down-glacier (Figure S.23). Sichelwannen are a type of **p-form**.

FIGURE S.23   A sichelwanne on the top of a roche moutonnée at Columbia Glacier, Prince William Sound, Alaska. Icebergs from the rapidly disintegrating tongue of this tide-water glacier are in the background. Glacier flow was towards the people in the photograph (July 1989).

S

Sikussak (derived from Greenlandic; synonym **ice mélange**)

A dense floating mass of ice in front of calving tidewater glaciers, comprising a mixture of icebergs, bergy bits and sea ice, partially welded together as an unstable mass of dense ice (Figure S.24).

FIGURE S.24   Kangerdlugssuaq Gletscher (left) is the largest outlet glacier from the Greenland Ice Sheet on the east coast of the Greenland. Icebergs, bergy bits and sea ice in front of this glacier coalesce to form an extensive zone of sikussak (May 2006, seen from a Transatlantic flight).

Sill (synonym **Threshold**)

A submarine barrier of rock or moraine that occurs at the mouth of a fjord (Figure S.25). A sill occurs sea-wards of an **overdeepened basin** where glacial erosion is inhibited by glacier thinning and declining velocity. In contrast, the overdeepened basin is the zone of maximum erosion. Some sills are associated with rocky islands. A sill may also occur also between two rock basins within the same fjord.

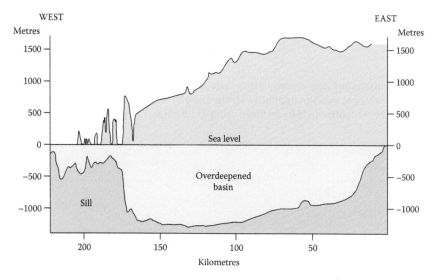

FIGURE S.25   A longitudinal profile through Norway's longest fjord, Sognefjord, showing a complex sill at the mouth (left), an overdeepened basin behind (middle) and the morphology of the side of the fjord above water-level. (Simplified from Holtedahl, H. 1967. Notes on the formation of fjords and fjord-valleys. *Geografiska Annaler*, 49, 188–203.).

Similar fold

A **fold** that is the result of ductile deformation, with a thickened hinge and attenuated limbs (see Figure F.24 for fold terminology). Similar folds are the commonest type of folds in glaciers (Figure S.26).

FIGURE S.26   A similar fold in Gulkana Glacier, Alaska Range, Alaska, illustrating thickening of layers at the fold hinge, just left of the lens cap (June 2012).

Simple shear

Deformation of a material by displacement on discrete (often closely spaced) surface or shear planes. The orientations of the principal axes of **cumulative strain** rotate as deformation proceeds (i.e. the strain is non-coaxial). Simple shear is commonly demonstrated using strain ellipses (Figure S.27).

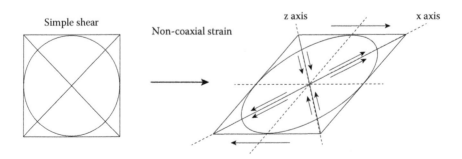

FIGURE S.27   Schematic illustration of simple shear in relation to the principal axes of strain, showing deformation of a circle into an ellipse.

Sliding (see **Basal sliding**)

Slope failure (see **Slumping** and **Rock-slope failure**)

Slot gorge

A steep-sided gorge in bedrock, deeper than it is wide, commonly with vertical and overhanging walls. A slot gorge forms beneath a temperate glacier as a result of the erosive action of subglacial meltwater under high pressure (Figure S.28).

FIGURE S.28   Slot gorge cut in limestone by meltwater beneath Unterer Grindelwaldgletscher, Switzerland. The glacier covered this site as recently as the Little Ice Age.

Slow surge

A type of unstable glacier flow that is similar to a 'normal' **surge** (which are completed within a few years), but which take place on an order of magnitude longer time scale. A slow surge is characterised by transfer of ice from a **reservoir area** to a **receiving area**. During the surge a bulge (the surge front) moves through the **glacier**, which, on reaching the **snout**, leads to an advance (Figure S.29). Slow surges are known only to occur in **polythermal glaciers**.

FIGURE S.29 The most closely monitored slow surge was on Trapridge Glacier, Yukon, illustrated here just after the surge front reached the snout and led to advance of a steep frontal cliff. The most active phase lasted from about 1980 to 2000, with a peak velocity of 42 m/year, after which the surge velocities slowly declined to less than 9 m/year in 2005 when the surge ceased (July 2006).

Slumping

The process of downslope mass movement whereby the internal organisation of the mass of debris is not totally disaggregated. Deformation may be evident in the form of recumbent fold structures and faults. Large-scale slumping in glacial environments occurs when deep-seated failure occurs in a **lateral moraine** following downwasting of the adjacent glacier (Figure S.30).

FIGURE S.30 A large slump at the margin of Khumbu Glacier following downwasting from its Little Ice Age level. The slump is marked by the displacement downwards of the brown vegetated area (April 2003).

Slush flow (slush avalanche)

A fast-flowing mass of water-saturated snow (or its end product), most commonly occurring in early summer when melting is at its peak (Figure S.31).

FIGURE S.31 A freshly deposited slush flow on an unnamed tributary glacier of Polarisbreen, northwestern Spitsbergen, Svalbard, showing characteristic finger-like lobes formed as different parts of the flow lose their water content in the underlying snowpack (July 2010).

Slush zone

The part of a glacier at and above the snowline where the snow is saturated with water. The slush may only be visible at the bottom of the snowpack (Figure S.32). See also **facies (snow)**.

FIGURE S.32 Slush zone on Rusty Glacier, Yukon, Canada. Walking across the wet snow on ice reveals sediment-stained slush just below the surface (July 2006).

Snout (also known as **Toe** in North America)

The lower part of the ablation area of a valley glacier (Figure S.33).

FIGURE S.33   Snout of a small valley glacier in Suess Land, East Greenland (July 1984).

Snow

Precipitation of ice crystals from the atmosphere, most of which are star-shaped or have a complex hexagonal form (Figure S.34a). The branches are commonly interlocking and at temperatures above about −5°C, the crystals are clustered into snowflakes. Fresh snow is light and fluffy with a density of only 1/10 that of water, or 100 kg m^{-3} (Figure S.34b), but rapidly becomes denser with settling and partial melting. On glaciers, snow accumulations maintain air and water circulation during the summer as density increases to around 400 kg m^{-3}. Snow that survives more than one ablation season and has a higher density than this becomes **firn**.

FIGURE S.34   (a) Single snow crystal. The intricate hexagonal shape stands out because of specular reflection of direct sunlight, and was taken after a fresh snowfall near the McGill Arctic Research Station on Axel Heiberg Island (1976).   *(Continued)*

(a)

(b)

FIGURE S.34 (Continued) (b) Freshly fallen, low-density snow near the village of Roundthwaite, Cumbria, England.

Snow avalanche (see avalanche)

Snowball Earth

The name given to the theory that the Earth was almost completely covered by glacier ice in Late Precambrian (Neoproterozoic) time (1000 to 570 million years ago) (Figure S.35).

FIGURE S.35 A key geological section in arguments in favour of the Snowball Earth theory, with its chief protagonist, Paul Hoffman, for scale. The section is near Narachaamspos in the Kaokoveld, Namibia. It shows termination of Snowball Earth conditions, with a transition from glacial rocks (tillites) with ice-rafted drop-stones, to characteristic cap carbonates.

S

Snow bridge

An arch of snow that has drifted across a linear cavity in the glacier surface, such as a **supraglacial stream channel** (Figure S.36a) or **crevasse** (Figure S.36b). Snow bridges may completely cover the cavity, and as such are a danger to travellers, especially when light conditions are too poor to pick out surface irregularities.

(a)

(b)

FIGURE S.36 Two types of snow bridge. (a) Partially collapsed snow bridge covering a deep meandering supraglacial stream channel on Midtre Lovénbreen, northwestern Spitsbergen, Svalbard (August 2009). (b) Aerial view of huge transverse crevasses with sagging snow bridges at the edge of the Polar Plateau, Antarctica, formed where the East Antarctic Ice Sheet begins to flow seawards *via* Shackleton Glacier. Some of these crevasses are estimated to be 50 m wide (January 1996).

Snow–ice avalanche pit

A pit created by the impact of an ice or snow **avalanche** onto soft ground. Small ridges of displaced material may form a rim. In fossil examples the floor of the pit is commonly marshy (Figure S.37).

FIGURE S.37   A circular ice-avalanche pit in Mosedale, southwestern Lake District, England. This was generated during the Younger Dryas glacial phase when an ice cap was situated on top of the hill in the background.

Snowline (or **snow line**)

(a) The altitude above which there is a continuous or near-continuous cover of snow (Figure S.38). (b) The **climatic snowline** is the altitude above which, on average, snow occurs throughout the year. Some examples of the latter are Svalbard, 300–600 m; Rocky Mountains, 3700–4000 m; the Alps, 2700–3200 m; Himalaya, 6000 m; tropical Andes, 5800–6500 m; Southern Alps (New Zealand), 1600–2700 m; Antarctica, 0–400 m.

FIGURE S.38   At the height of the Southern Hemisphere summer, the snowline on glacierised Osorno Volcano, Chile, is prominent. Whereas the summit is at an altitude of 2652 m.a.s.l., the snowline is around 2100 m (January 1996).

Snow stratigraphy

The analysis of the vertical structure of a snowpack to determine changes through time. On glaciers, determining snow stratigraphy is a crucial component of **mass-balance** studies. Typically, snow pits are dug to establish the water equivalent of snowfall each accumulation season (Figure S.39a) or, at the end of the ablation season, to establish what is left to contribute to glacier mass. Ice layers are recorded and density measurements taken to determine these attributes (Figure S.39b). Longer term records of snow stratigraphy, as in **ice cores**, are obtained to determine changes in climate over decades to hundreds of thousands of years.

(a)

FIGURE S.39   (a) Glaciologist Jemma Wadham acquiring a snow stratigraphic record from the accumulation area of a small valley glacier, Midtre Lovénbreen, northwestern Spitsbergen, Svalbard. The hand-dug pit is about 2m deep (April 1996).   *(Continued)*

(b)

FIGURE S.39 (Continued) (b) Holding a snow block from a pit against the sun illustrates the layered nature of snow, each layer being measured for structure and density. This photograph was taken at 2540 m.a.s.l. near Weissfluhjoch, Davos, Switzerland.

Snow swamp

An area of saturated snow lying on glacier ice. Snow swamps are especially widespread on polythermal glaciers early in the melt season. Such areas may not be immediately visible and are a trap for the unwary (Figure S.40).

FIGURE S.40 Snowmobile after an unsuccessful attempt to drive through a snow swamp on White Glacier, Axel Heiberg Island, Canada (1976).

Sole (see **glacier sole**)

Spalling

The breaking off of lumps of rock, sediment and ice from a steep face. The term is especially pertinent to the mass movement of **glacigenic sediment**, for example, **basal till**. Masses of sediment associated with a decaying glacier margin are particularly prone to mass movement as melting ice provides the lubricant to initiate flowage. Flow commonly begins by fracture of sediment to form small cliff faces, with blocks of sediment mixing with meltwater flowing down slope, and eventually becoming totally disaggregated (Figure S.41). Another form of spalling, affecting bedrock, is **exfoliation**.

FIGURE S.41 Basal till being subjected to spalling, with the debris flow generated from melting ice beneath moving towards the right. The till was deposited by the glacier in the background, Kongsbreen, northwestern Spitsbergen, Svalbard (July 2009).

Spillway

A distinctive type of meltwater channel, formed immediately in front of, and beyond, a glacier or ice-sheet margin. It is produced when meltwater in an ice-marginal lake overflows *via* a col or low point in a watershed. It can be cut over a long period of time by regular discharge from the glacier (Figure S.42).

FIGURE S.42 The spillway of Newtondale, North Yorkshire, England. It was cut by meltwater draining from a proglacial Lake Eskdale across unglaciated uplands to Lake Pickering in the south. The valley is followed by the heritage North Yorkshire Moors Railway.

Spindrift

Snow that is blown into the air during high wind (Figure S.43). This is a common feature in mountain and polar regions. The term is also used for sea spray.

FIGURE S.43 Spindrift caused by a strong northeasterly wind on Las Sours as seen from Muottas Muragl near Pontresina, Switzerland (February 2009).

Splaying crevasses

A set of **crevasses** that fan out down-glacier, especially where the glacier becomes wider (Figure S.44). The maximum tensile stress, acting perpendicular to the crevasses is at an angle to the valley sides. Splaying crevasses are particularly well developed in **piedmont glaciers**.

FIGURE S.44 Splaying crevasses fanning out below a set of transverse crevasses (top left), which they intersect, Vadret da Morteratsch, Switzerland. Telephotograph from Munt Pers (September 2012).

Spring event

A sudden drainage event from an alpine glacier in late spring. It follows a period of rapid melting or heavy rain. The water level in the glacier is high, commonly reaching the surface, until basal conduits open sufficiently for rapid discharge to take place. Hydraulic jacking may be implicated in the event. The outflow has a high sediment load.

Stadial (or stade)

A short-lived period of cold, commonly associated with glacier growth. The term is not precisely defined, however. It is most widely used for subdividing the latter stages of the last glacial period (the Wisconsinan in North America, the Weichselian in northern Europe, the Würm in the Alps and the Devensian in Britain). The Last Glacial Maximum was followed by a series of stadials and interstadials, of which the best known is the **Younger Dryas Stadial**, lasting from c. 13,000 to 11,500 years BP.

Stagnant ice (synonym of Dead ice)

Any part of a glacier or ice sheet that does not flow at a detectable rate. Dead ice is commonly associated with a cover of **supraglacial debris** and glaciotectonically entrained **basal debris**. Areas of stagnant ice with well-developed ponds, channels and depressions are referred to as **glacier karst**. **Surge-type glaciers** in their quiescent state are characterised by extensive areas of dead ice (Figure S.45).

FIGURE S.45 Aerial view of an area of stagnant ice that underlies a wide belt of lateral moraines near the snout of Sermilik Glacier, Bylot Island, Canadian Arctic (July 2014).

Stapi (from Icelandic; see **Tuya**)

Sticky spot

Areas beneath an **ice stream** that have anomalously high friction, thus locally slowing down the fast-flowing ice. Sticky spots result from bedrock bumps, areas free of till, areas of stiff (well-drained) till and freeze-on of subglacial meltwater. Since ice streams commonly produce flow-parallel **streamlined bedforms**, flow will be diverted around the sticky spot and the landforms will locally change orientation.

Stoss-and-lee form

A bedrock or boulder feature that has a smooth (commonly striated) up-glacier slope (stoss side) and a rough, steep down-glacier (lee) side. Whereas the term **roche moutonnée** is used for solid bedrock forms, stoss-and-lee forms are applied to large boulders that are embedded in substrates such as **till** (Figure S.46).

FIGURE S.46 A large stoss-and-lee boulder embedded in the lateral moraine of Ghiacciaio dei Forni, in the Ortles-Cevedale region of the Italian Alps (August 2008).

Strain

The amount by which an object becomes deformed as a result of stress, expressed either in change of length or shape (in both two and three dimensions) (Figure S.47).

FIGURE S.47 Sean Fitzsimons measuring the amount of strain in two dimensions in the basal ice layer of Taylor Glacier, Victoria Land, Antarctica (January 1999).

S

Strain ellipse

The ellipse that results from the deformation of a sphere, either theoretically or naturally. The strain ellipse has two principal axes that define the directions of maximum and minimum strains (Figure S.48). The strain ellipsoid is the three-dimensional equivalent, additionally having an intermediate strain axis. Strain ellipses or **strain markers** are commonly represented by water-filled **moulins** on glaciers.

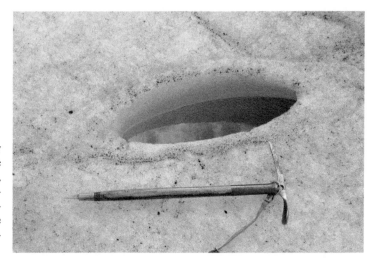

FIGURE S.48 A naturally occurring strain ellipse at the surface of Vadret da Morteratsch, Switzerland. These quite common features were original spherical water-filled moulins that have deformed in a simple shear deformation regime.

Strain net

An arrangement of pegs or stakes in a glacier to determine rates of strain (Figure S.49). The distances between pegs and markers are typically measured over periods of a month to a year, and the data are used to calculate the principal strain-rate axes.

FIGURE S.49 Small white pegs, about 20 cm apart, originally arranged in a square have deformed by simple shear into a parallelogram over the course of a year, in the basal ice layer of Taylor Glacier, Victoria Land, Antarctica (*cf.* Figure S.47) (January 1999).

Strandline (see **lake shoreline**)

Stratification

The annual layering that forms from the accumulation of snow on a glacier, and which is preserved in **firn** and in **glacier ice** as continuous layers of coarse bubbly ice, with coarse clear/dirty ice representing the summer surface (Figure S.50). During glacier flow, stratification is subject to folding and over-printing by other structures.

FIGURE S.50 Stratification cropping out at the surface of a branch of Trapridge Glacier, Icefield Ranges, Yukon. The individual layers are about 0.5 m thick on average (July 2006).

Streaklines (see **flowlines**)

Streamlined bedforms/topography

The collection of landforms with a smooth, rounded and elongated morphology, which are the result of fast glacier flow, such as beneath an **ice stream** or **valley glacier** (Figure S.51).

FIGURE S.51 Streamlined bed-forms, represented by the green fields, trending left to right in the Vale of Threlkeld, near Keswick, English Lake District. The sub-surface is till and provides good pasture for sheep and cattle.

S

Stress

The measure of force per unit area, which represents how hard a material (such as glacier ice) is being stretched, compressed or twisted in response to an external force. The unit of stress is Newtons per square metre or Pascals. For the high stresses that occur, for example, at the base of a glacier, kilopascals are commonly used (1 kPa = 1000 Pa). In older literature the units used were bars (where 1 bar = 100 kPa). Various stresses can be defined. Stresses on a surface have two components: the **normal stress** acting at right angles to the surface and the **shear stress** acting parallel to the surface. Normal stresses have two components: a **tensile stress**, which is the result of stretching and conventionally has positive values, and **compressive stress** where the material is being squeezed and has negative values. The term **yield stress** refers to the relationship between strain rate and shear stress, whereby material such as ice does not deform until a critical shear stress is reached, at which point the material begins to flow. This relationship is illustrated in Figure P.12.

Striae or striations

Linear, fine regular scratches or small grooves formed by the abrasive effect of debris-rich glacier ice sliding over bedrock (Figure S.52a). The process is not smooth, but jerky, generating microfractures, so striae record the combined effect of **abrasion** (which results in polished rock) and brittle failure. Intersecting sets of striae are formed as stones are rotated within the ice, or if the direction of ice-flow over bedrock changes. Striae occur on both bedrock and the stones that have been transported at the base of a glacier (Figure S.52b). Some striae begin faint and shallow, then deepen and end abruptly (nail-head striations), others have miniscule **chattermarks** in the base of the scratch mark. Striations may also form beneath cold ice, but these are rougher and irregular in appearance (Figure S.52c). **Striated** is the term for the scratched state of bedrock or stone surfaces after the ice has moved over them.

(a)

(b)

FIGURE S.52 Striations on bedrock and basally transported clasts. (a) A recently exposed striated bedrock (granite) surface, with a scattering of erratics, near the margin of the temperate glacier Gornergletscher, Switzerland. The peak with the small cloud is the Matterhorn. (b) Striated limestone boulder from within basal till near the snout of the polythermal glacier Pedersenbreen, northwestern Spitsbergen, Svalbard.   *(Continued)*

(c)

FIGURE S.52 (Continued)   (c) Rough striations on a gneiss boulder at the edge of an unnamed cold-based glacier in southern Bylot Island, Canadian Arctic.

Stromatolites

Mats, mounds and columns of blue-green algae together with trapped sediment. They are a feature of cold arid environments, such as the Dry Valleys, Antarctica, where they occur in shallow lakes next to **cold glaciers** (Figure S.53). This modern example has been regarded as a suitable analogue for the abundant stromatolites that occur in association with late Precambrian **tillites**.

FIGURE S.53   Small stromatolite mounds poking through the permanent ice surface of Lake Brownworth, an ice-contact lake adjacent to Wright Lower Glacier, Victoria Land, Antarctica. These stromatolites are up to a metre long and grow upwards from the floor of the lake. They have the texture of partially rotted cardboard.

Subaqueous sediment fan (see **grounding-line fan**)

Subglacial cavity

An open space formed at the base of a sliding glacier, especially where a temperate glacier flows over a bedrock step or bump (Figure S.54). In this zone, freeze–thaw, **regelation** and sediment entrainment are important processes.

FIGURE S.54   Cavity beneath Glacier de Tsanfleuron, Switzerland, with Bryn Hubbard examining regelation ice layers and folding (August 2000).

Subglacial channel/conduit

A drainage tunnel at the bed of a glacier, either incised into the underlying rock or sediment (Figure S.55a), or upwards into the ice (Figure S.55b), by meltwater. See also **Nye channel** and **Röthlisberger channel,** respectively. Channels in bedrock are valleys with steep, commonly vertical sides. They are eroded by a subglacial stream under high pressure. Such channels may have a profile with uphill sections, since a sufficient head of water in the glacier can force it to flow up-slope. The best-developed examples of subglacial channels are referred to as **slot gorges**.

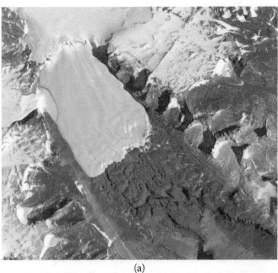

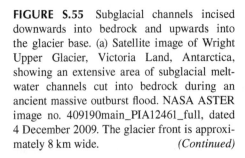

FIGURE S.55 Subglacial channels incised downwards into bedrock and upwards into the glacier base. (a) Satellite image of Wright Upper Glacier, Victoria Land, Antarctica, showing an extensive area of subglacial meltwater channels cut into bedrock during an ancient massive outburst flood. NASA ASTER image no. 409190main_PIA12461_full, dated 4 December 2009. The glacier front is approximately 8 km wide.   *(Continued)*

(a)

(b)

FIGURE S.55 (Continued) (b) Channel incised into the base of Stagnation Glacier, Bylot Island, Canadian Arctic, by a subglacial stream at the right-lateral margin (July 2014).

Subglacial debris

Debris that has been released from ice at the base of a **glacier** by a range of processes including **melt-out** and **lodgement** (Figure S.56). Individual stones usually show signs of rounding as a result of **abrasion** at the contact between ice and bedrock. The process of debris release is referred to as subglacial deposition (*cf.* **basal debris**, which is debris still within the ice).

FIGURE S.56 Subglacial debris beneath Midtre Lovénbreen, northwestern Spitsbergen, Svalbard. The sediment is frozen and has a sharp contact with overlying clean glacier ice that forms the roof of the tunnel (August 2009).

Subglacial gorge (see **slot gorge**)

Subglacial hydrology (or **basal hydrology**)

The study of meltwater-related processes at the base of glaciers, especially where meltwater gathers into distinct **subglacial streams** or forms **subglacial lakes** (Figure S.57). Subglacial hydrology is important for assessing flood potential, the contribution of meltwater to hydroelectric power schemes and irrigation. Subglacial hydrology also plays an important, but little-studied role in the net **mass balance** of a glacier, through removal of ice by erosion.

S

FIGURE S.57 Subglacial hydrology of temperate alpine glaciers is reflected in the opening of large tunnels and caves at the base, and their subsequent collapse, as here at Vadret da Morteratsch, Switzerland (May 2010).

Subglacial lake

A lake that is formed beneath a **glacier** or **ice sheet** where ice is at the pressure melting point. The lake typically fills an overdeepened basin and can connect with others *via* **subglacial streams**. Abundant lakes have been detected beneath the Antarctic Ice Sheet where geothermal heat warms the base (Figure S.58). Such lakes are at the frontier of exploration, but are considered to be host to unique ancient ecosystems.

FIGURE S.58 Map of Antarctica showing the location of all known lakes detected by geophysical methods. The colours and shapes indicate the type of investigations undertaken at each site: radio-echo sounding – black/triangle; seismic sounding – yellow; gravitational field mapping – green; surface height change measurement – red/circle; shape identified from ice surface feature – square. The largest of these lakes, Vostok, is shown in outline. (Map from Wright, A. and Siegert, M. 2012. A fourth inventory of Antarctic subglacial lakes. *Antarctic Science* 24, 659–664.)

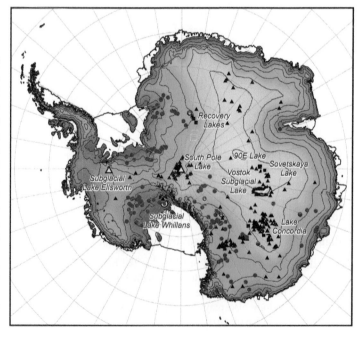

Subglacial meltwater channel (see **subglacial channel/conduit**)

Subglacial sediment (see **subglacial debris**)

Subglacial stream (see **subglacial hydrology**)

Subglacial till mosaic

The complex suite of sediments that is produced beneath a glacier. The range of processes beneath a **temperate** or **polythermal glacier** includes **melt-out**, **lodgement**, deformation, flow, sliding and ploughing. These processes coexist and act to mobilise and transport sediment, resulting in sediments that can be classified descriptively (e.g. stratified sediment, **diamicton**), but less easily using traditional genetic terms such as melt-out **till**, lodgement till or deformation till.

Subglacial traction till

An over-arching term for all massive (non-bedded) subglacial tills. The term replaces single-process terms such as melt-out till, lodgement till or deformation till, because it is now recognised that practically all subglacial tills are hybrids of all of the above processes, but with varying degrees of importance (Figure S.59).

FIGURE S.59   Sarah Trueman inspect a subglacial traction till from the Late Glacial Maximum at Tràigh na h-Uidhe, South Harris, Outer Hebrides, Scotland. This deposit illustrates a combination of processes including lodgement, meltout and possibly glaciotectonic deformation.

Subglacial volcanism (see also **Glaciovolcanism; Tuya**)

The eruption of a volcano beneath a **glacier** or **ice sheet** (Figure S.60a). The interaction of magma with ice generates huge volumes of meltwater, and the resulting outburst flood is known as a **jökulhlaup**. Subglacial eruptions also produce a distinctive sequence of volcanic rocks, including lava, **tephra** and **hyaloclastite**, sometimes built up into substantial mountain landforms (Figure S.60b). Layers of tephra in glacier ice may be used to track ice dynamic behaviour. Detailed study of glaciovolcanic sequences yields information about ice thickness, the basal **thermal regime**, the elevation of the ice surface and the structure of the ice mass. Obtaining a chronology of events is aided by the ability to accurately date volcanic rocks radiometrically spanning millions of years.

(a)

(b)

FIGURE S.60 Subglacial volcanism processes and products in Iceland. (a) The 2010 eruption of Eyjafjallajökull, which caused havoc to European aviation (photograph by Marco Fulle). (b) Öræfajökull, a glacier-covered volcano in southeast Iceland, reaching over 2000 m in height. It generated jökulhlaups in 1362 and 1727, and produced large amount of tephra. Tephra layers are present in the surrounding glaciers and in the icebergs derived from them (March 2015).

Subglacial water pressure (see **upwelling; fountain**)

Sublimation

The process whereby ice changes to vapour without melting. When this process takes place in debris-rich ice, the resulting deposit is referred to as **sublimation till** (Figure S.61). Wind-scoured glacier surfaces in Antarctica are prone to sublimation and produce **blue ice areas**.

FIGURE S.61 Sublimation till forming near the snout of Taylor Glacier, Victoria Land, Antarctica. The fawn deposit at the top is the till, and the grey mass below is debris-rich basal ice from which the till is derived (November 1986).

Subpolar glacier

An obsolete term for a glacier, the ice of which is partly sub-zero and partly at the pressure melting point. This term is geographically misleading, and the term **polythermal glacier** is preferred instead.

Summer surface

The surface formed when there is minimum mass across a whole glacier at the end of the ablation season. The surface is diachronous; that is, it varies in time for different parts of a glacier with the upper part of the glacier starting to gain mass whilst the lower part is still ablating. The summer surface is marked by relatively dirty (dust-covered) glacier ice, a superimposed ice layer and dirty snow with a part-melted crust (Figure S.62).

S

FIGURE S.62 The summer surface of Vadret da Tschierva, Switzerland, near the end of the ablation season. New snow at high elevation (approximately 4000 m) heralds the start of the accumulation season, but ablation continues over the grey ice below. The intervening orange-stained snow between is Saharan dust, highlighting the summer surface in the lower accumulation area (September 2014).

Suncups (or sun-cups)

Small bowl-shaped interlocking depressions in an ablating snowpack, giving rise to a honeycombed appearance. The bowls are typically 5–20 cm in diameter and a few centimetres deep, with sharp ridges between. In clean snow they form under conditions of intense solar radiation (Figure S.63), but on dirty snow, local turbulence of wind can generate similar features (*cf.* **scalloped ice**).

FIGURE S.63 Suncups formed in a snowbank at the edge of McMurdo Ice Shelf, Minna Bluff, Antarctica. Mount Discovery is in the background (December 2010).

Supercooling (see adfreezing)

Superimposed ice

Ice that forms on the surface of a glacier as a result of the freezing of rainwater or water-saturated snow. It is typically opaque and contains air bubbles, although fewer than in glacier ice derived directly from snow. Superimposed ice may cover large parts of a glacier, including (a) the ablation area where it forms at the base of the winter snowpack, but disappears as the ablation season progresses (Figure S.64) and (b) the lower accumulation area where it contributes to the gain in mass during the winter season. Superimposed ice is best developed on **polythermal glaciers**, on some of which it represents the principal contribution to accumulation in each **mass-balance** year. See also **facies (snow)**.

FIGURE S.64 Superimposed ice formed at the base of the winter snowpack in the ablation area of Austre Lovénbreen, north-western Spitsbergen, Svalbard. This ice, which was several centimetres thick, disappeared after a few days of melting, revealing the rough glacier ice beneath (July 2013).

Supraglacial debris

Debris that is carried on the surface of a glacier. It is normally derived from rockfalls and tends to be angular in character, with particle sizes ranging from sand to boulders (Figure S.65). Some supraglacial debris, particularly towards the **snout** of a glacier, may be derived from the glacier bed, for example, by **thrust-faulting**. This debris has the attributes of **basal debris**.

S

FIGURE S.65 Supraglacial debris, derived from rock-fall, on the Oberaargletscher, Switzerland. Angular cobbles and boulders are typical and provide the bulk of material in this medial moraine. The mountain in the background is Oberaarhorn (September 2009).

Supraglacial debris stripe

A distinctive linear trail of angular debris that stretches across the proglacial area of a glacier. It is derived from **supraglacial debris** that was buried in the upper reaches of a glacier, and then folded with axes parallel to flow. The folded debris re-emerges as a **medial moraine** or **debris septum** in the **ablation area**, and is deposited on top of basally deposited sediments as the glacier recedes (Figure S.66). The orientation of a group of debris stripes thus reflects both the original structure within the glacier and direction of flow.

FIGURE S.66 Supraglacial debris stripes on the proglacial area of Midtre Lovénbreen, northwestern Spitsbergen, Svalbard. The stripes are denoted by the low darker ridges of angular material, and they are superimposed on pale-coloured subglacial till (July 1995).

Supraglacial drainage

Drainage on the surface of a glacier or **ice sheet**. This takes many forms including small channels (**rills**), supraglacial streams of straight or meandering form, **cryoconite holes** and **supraglacial lakes** and ponds (Figure S.67). Channels commonly disappear into **moulins** and water joins an englacial and subglacial drainage system. The overall drainage system is thus three-dimensional.

FIGURE S.67 The complexity of drainage is illustrated by this deeply incised, meandering supraglacial stream and associated pond, as well as rills, cryoconite holes, and scattered supraglacial debris on Gornergletscher, Switzerland (August 2011).

Supraglacial lake/pond

A lake or pond formed on the surface of a glacier. The larger lakes tend to form on **polythermal glaciers**, as drainage is impeded by ice that is below the pressure melting point. Lakes and ponds commonly have sediment and biogenic material (**cryoconite**) within them, thus absorbing more solar radiation and thereby enhancing their depth (Figure S.68).

FIGURE S.68 A lake about 80 m across on the surface of Gornergletscher, Valais, Switzerland. This glacier is one of the few polythermal glaciers in the Alps, the ice originating from high altitude. The white ice contrasts strongly with the blue lake with its cryoconite-covered bottom (August 2011).

Supraglacial stream

A stream that flows over the surface of a glacier. Most supraglacial streams descend *via* **moulins** into the depths or base of a glacier. Supraglacial streams may run for many kilometres on large **polythermal** or **cold glaciers**. Their channels may be straight, irregular or meandering, and incised to depths of many metres (Figure S.69a). As in rivers, meander cut-offs may form. Supraglacial streams are influenced by structures such as **crevasses** and **foliation** (Figure S.69b). If the ice is dynamic, channels may close during the winter as a result of ice **creep**. In exceptional cases supraglacial channels may grow into **canyons**.

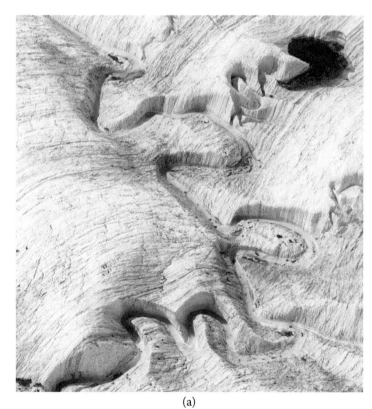

(a)

FIGURE S.69 Examples of supraglacial streams. (a) Aerial view of meandering stream on Root Glacier, Wrangell Mountains, Alaska (July 2011).

(Continued)

S

(b)

FIGURE S.69 (Continued) (b) The straight reach of a supraglacial stream on Auste Lovénbreen, northwestern Spitsbergen, Svalbard; this stream has formed parallel to strong vertical longitudinal foliation (August 2013).

Surge

A short-lived phase of accelerated glacier flow during which the surface becomes broken up into a maze of **crevasses** (Figure S.70). A surge may last a few months or several years and may end abruptly. Surges are often periodic and are separated by longer periods of relative inactivity or even stagnation, referred to as **quiescence**. The shortest quiescent phases last just over a decade, but most last for several decades. Velocities in a **surging glacier** are at least an order of magnitude greater than those during the quiescent phase. During a surge, ice is transferred from a **reservoir area** to a **receiving area**, sometimes with a bulge (representing a **kinematic wave**) moving rapidly through the glacier. As a result, the upper part of the glacier undergoes surface-lowering and the lower part thickening. Surges do not occur in all **glacierised** areas. Prominent regions with surge-type glaciers include Alaska/Yukon, the Canadian Arctic and Svalbard; regions without surge-type glaciers include the Alps, mainland Norway and (as far as is known) Antarctica. Surges are related to changes in the basal hydrological regime and are not directly related to climatic change (see also **slow surge**).

FIGURE S.70 The surge of Comfortlessbreen, northwestern Spitsbergen, Svalbard, in July 2009. The surge ceased in 2010 after lasting about 5 years, but the glacier had no prior record of surging, so the length of its quiescent period is unknown. In this image the surge is pushing up a ridge of marine sediment, and major modifications of the proglacial area are underway.

Surge front

The zone of intense compression between surging ice and non-surging ice. This is commonly marked by a bulge, and a transition from heavily crevassed to crevasse-free ice. The surge front rapidly moves through the glacier, and if it reaches the snout, the glacier advances (Figure S.71).

S

FIGURE S.71 The surge front of Trapridge Glacier, Yukon, in July 2006, about a year after it reached the snout and resulted in a small advance. The surge front took over 20 years to pass through this polythermal glacier, and the phenomenon is referred to as a slow surge (see also Figure S.29).

Surge loops (see **looped moraine**)

Surge moraine

An irregular **end-moraine** formed as a result of a rapid glacier advance during a surge, followed by ice stagnation. The moraine is hummocky with few well-defined ridges and comprises a mixture of basally and supraglacially transported material. Extensive areas **of dead ice** underlie this debris, which melts only slowly under the thick debris mantle (Figure S.72).

FIGURE S.72 Aerial view of the surge moraine at the terminus of Muldrow Glacier, Alaska Range, Alaska, with a flat braidplain beyond. The glacier, which originates on Denali (formerly Mt McKinley), last surged in 1956–1957 (June 2012).

Surge-type glacier

A glacier that periodically undergoes rapid flow, sometimes with advance of the terminus (a **surge**), separated by a longer period of low flow or stagnation (**quiescence**) (Figure S.73). See **surge** for further details.

FIGURE S.73 A temperate surge-type glacier in its quiescent state: Lowell Glacier, Icefield Ranges, Yukon, in July 2006. The glacier has impounded the Alsek River, creating a large ice-dammed lake. Note also the distorted nature of the medial moraines.

S

Surging glacier

A glacier that is actively undergoing a **surge** (Figure S.74).

FIGURE S.74 The advancing front of an actively surging poly-thermal glacier: Frijthovbreen, western Spitsbergen, Svalbard, in July 1996. In this state the front is totally broken up and is advancing at several metres a day. Compare with Figure S.70.

Suspended sediment

Sediment of **silt** and **clay** particle size that is carried in suspension in fast-flowing currents. In glacial environments, suspended sediment is typically carried by glacial streams (Figure S.75) or by currents where sub-glacial streams enter a lake or the sea (see also **sediment plume**).

FIGURE S.75 Convergence of two supraglacial streams with suspended sediment, reflecting different geological sources at Austre Brøggerbreen, north-western Spitsbergen, Svalbard. The reddish stream contains sediment from red sedimentary rocks of Devonian age, and the greenish grey stream is from Proterozoic metamorphic rocks (August 2009).

S

Syncline

A downfold of layers in a geological material, with the youngest layers on top. Many glaciers have a syn-clinal structure when seen in cross-section, especially where they terminate as a calving cliff in lakes. The syncline may comprise the original **stratification** or involve deformation of later structures such as **arcuate foliation** (Figure S.76).

FIGURE S.76   Syncline stretching across the full width of debris-mantled Hooker Glacier, South Island, New Zealand. The synclinal structure is enhanced by layers of debris, derived from periodic rockfalls into the accumulation area of the glacier, from adjacent mountains such as Aoraki/Mt Cook (April 2008).

S

T

Tabular berg

A flat-topped **iceberg** that has become detached from an **ice-shelf**, **ice tongue** or floating **tidewater glacier**. Tabular bergs are typically several kilometres long, and several hundred metres thick (Figure T.1). Antarctic tabular bergs may attain dimensions of over 100 km, and in a large group may be referred to as an **iceberg armada**.

FIGURE T.1 Tabular iceberg off the northeastern Antarctic Peninsula, showing an elevated waterline as a result of tilting resulting from uneven melting underwater.

Talus (synonym: scree)

Accumulations of loose rockfall debris cone on steep slopes below cliffs, typically on recently deglaciated terrain. As a result of freeze–thaw processes, talus comprises coarse angular boulders, cobbles, pebbles and a little sand. Until stabilised by vegetation, the deposit is loose and contains large void spaces. A **talus cone** is a fan-shaped accumulation of rockfall debris below a gully (Figure T.2a). Cones may coalesce to form an extensive **talus apron** on a mountainside. Talus may also form from a steep-sided **moraine**, in which case, the textural characteristics resemble those of the parent body. Talus may be modified by avalanching in spring when snow, meltwater and debris mix to form a sediment flow (Figure T.2b).

(a)

(b)

FIGURE T.2 (a) A talus cone along the flanks of Lake Louise, Banff National Park, Canada. The turquoise colour of the lake is from suspended sediment carried by a stream from a nearby glacier. (b) Snow avalanche-related flow lobes on a talus slope in the Hooker Valley, Aoraki/Mt Cook National Park, South Island, New Zealand.

Tarn (a northern English term)

A small lake occupying a hollow eroded out by a glacier or dammed by a moraine (Figure T.3). Tarns are commonly but not exclusively associated with cirques.

FIGURE T.3 Sprinkling Tarn in the English Lake District, occupying a glacially scoured rock basin, overlooked by the shadowed peak of Great Gable.

Teardrop moraines (see **looped moraines**)

Temperate glacier (synonym: warm glacier)

A **glacier** whose temperature is at the pressure melting point throughout, except for a cold wave of limited penetration (approx. 10 m) that occurs in winter. A temperate glacier is **wet-based**, and commonly debris-mantled and having a **subglacial stream** emerging at the **snout** (Figure T.4a). Many temperate glaciers extend below the tree-line (Figure T.4b). The term relates to **thermal regime** and complements the terms **cold glacier** and **polythermal glacier.** Ice that is at the pressure melting point and intimately associated with meltwater is called **temperate ice**.

(a)

FIGURE T.4 Contrasting temperate glaciers. (a) The partially debris-mantled temperate Pasterzenkees, the largest glacier in the Austrian Alps, showing emergence of meltwater at its decaying snout. Note many tourists approaching and hiking towards and on to the glacier tongue (August 2009).

(Continued)

FIGURE T.4 (Continued) (b) The clean Rob Roy Glacier, South Island, New Zealand, framed by the branches of a southern beech and emphasising the close association of a temperate glacier with forest (January 2011).

(b)

Tephra

Ash and coarse fragments thrown out by a volcano during an eruption (Figure T.5). In glacierised areas these layers are preserved in the glaciers, deforming slowly during ice flow.

FIGURE T.5 Ash from the 1995 eruption of Ruapehu, North Island, New Zealand, covers the surface of Plateau Glacier, which fills a dormant crater. Harry Keyes peers into one of the many pits that result from ice melting unevenly beneath the tephra layer (April 1996).

Terminal moraine (synonym: end moraine)

The outermost ridge of glacial debris formed at the limit of a glacier advance (Figure T.6). This general term has no genetic connotations but embraces moraine types defined by process, such as a thrust moraine, push moraine or dump moraine.

FIGURE T.6 Terminal moraine of Uvêrsbreen, a polythermal valley glacier in northwestern Spitsbergen, Svalbard. This moraine dates primarily from the Neoglacial period of around 1900, but also includes some older elements (July 1992).

Terminus (pl. termini)

The lowest or outermost end of a **glacier**. It has a similar meaning to snout or toe but is most frequently applied to **tidewater** or **freshwater glaciers** (Figure T.7).

FIGURE T.7 Terminus of Nordenskiöldbreen, a grounded tidewater glacier in Isfjorden, Spitsbergen, Svalbard. The sea ice in the foreground (with an early summer melt-pool) has been pushed up slightly by a winter readvance (July 1981).

Thermal regime

That state of a glacier as determined by its temperature distribution. See **temperate glacier**, **cold glacier** and **polythermal glacier**.

Thermal stress fracture

A weathering process resulting from the fracturing of rock by stress fatigue and shock induced by temperature changes or temperature contrasts (*cf.* **frost action**). The process affects bedrock and boulders and is particularly prevalent in cold arid regions such as Antarctica. The effect of the sun on boulders overcomes air temperatures, and strong thermal gradients may develop in the rock, thus facilitating fracture. Fracturing may be deep-seated or result in surface flaking (Figure T.8).

FIGURE T.8 A large (3 m high) granite boulder in lower Taylor Valley, Victoria Land, Antarctica, that has large, deeply penetrating cracks, probably the result of thermal stress fracturing.

Threshold (see sill)

Thrust (ice)

A low-angle fault, usually formed where the ice is under compression. Thrusts commonly extend from the bed and are associated with debris and overturned **folds**. Thrusts carrying abundant debris are particularly common in polythermal glaciers (Figure T.9a). Thrusts have been observed actually forming in **surging glaciers**. In **temperate glaciers**, thrusting may commonly be observed at the snout, but generally the amount of debris is small (Figure T.9b).

FIGURE T.9 Examples of thrusting in valley glaciers. (a) A side-on view of thrusting, with entrainment of wedges of basal glacial sediment in Hessbreen, west-central Spitsbergen, Svalbard (August 1995). *(Continued)*

(a)

(b)

FIGURE T.9 (Continued) (b) The grounded terminus of temperate Cascade Glacier in Harriman Fjord, Chugach Mountains, Alaska. A series of thrusts drag small amounts of debris upwards on an upwardly curving trajectory (July 1989).

Thrust-block moraine

A **moraine** formed as an advancing **glacier** impinges upon frozen sediment, resulting in brittle failure along low-angle planes (thrusts), with stacking of one slice of sediment upon another (Figure T.10). Thrust-block moraines are high-Arctic and Antarctic phenomena, associated with **polythermal glaciers**. Proglacial outwash plains of gravel are especially prone to this form of **glaciotectonic deformation**.

FIGURE T.10 Thrust-block moraine in front of Thompson Glacier, Axel Heiberg Island, Canadian Arctic. After advancing since observations began on a regular basis in 1960, the glacier front has become stationary in recent years.

Thrust moraine

Related to **thrust-block moraine**, but involving both ductile and brittle deformation, within the glacier and in the sediment in front. Sediment is folded initially within the glacier, followed by shearing off of the lower limb of the **fold**. The resulting moraine includes englacial and proglacial components (Figure T.11a and b). The latter represent deformation in front of the glacier, and as such, form beyond the maximum advance of the glacier itself.

(a)

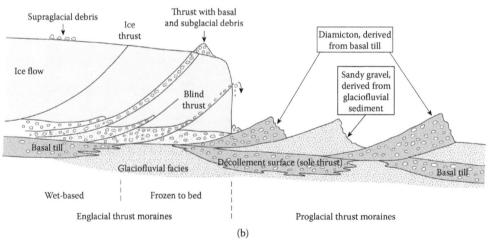

(b)

FIGURE T.11 (a) Thrust moraine complex, still with an ice core. A characteristic rectilinear slope faces the glacier, which lies off the photograph to the right. This moraine was formed during the 1948 surge of Kongsvegen, a polythermal glacier in northwestern Spitsbergen, Svalbard. The moraine includes glaciomarine sediment and is here pushed up onto Ossian Sarsfjellet (July 1996). (b) Conceptual model showing the development of thrust moraines of the Svalbard type in a longitudinal cross-section. Thrusting occurs both englacially and proglacially, the former facilitated by the transition from a sliding to a frozen bed. Stresses are propagated from the glacier into frozen sediments beyond, resulting in brittle failure.

Tidecrack

The fissure that occurs at the boundary between an immoveable **ice foot** or **ice wall** and tidally influenced fast ice (Figure T.12).

FIGURE T.12   A 2-m-wide tidecrack at the terminus of Mackay Glacier, western Ross Sea, Antarctica. The active sea ice on the right rises and falls with the tide, breaking off chunks of ice when the two sides of the fissure collide (December 1997).

Tidewater glacier

A glacier that terminates in the sea, usually in a fjord or bay. The term usually applies to glaciers with termini that are grounded on the sea bottom, although it is also sometimes used for floating glacier tongues (Figure T.13a and b).

(a)

FIGURE T.13   (a) Aerial view of calving Iceberg Glacier, Axel Heiberg Island, Canadian Arctic. The glacier terminus is partly floating, producing some small tabular bergs (July 2008).   *(Continued)*

T

(b)

FIGURE T.13 (Continued)   (b) The calving grounded terminus of Blomstrandbreen, northwestern Spitsbergen, Svalbard, shortly after the glacier surged (July 2013).

Till (an old Scottish word for 'a kind of coarse obdurate land')

Till has various definitions, but there is consensus that it is a sediment deposited directly by glacier ice, without subsequent disaggregation, reworking or resedimentation. Till is a genetic term, with the emphasis on process of deposition. Texturally, till is a highly variable sediment, but normally has a mixture of mud, sand and gravel in varying proportions. At one end of the spectrum it consists of mud, with minor proportions of sand and gravel; at the other, it is a sandy gravel dominated by boulders. The principal types of till are **basal**, deposited beneath a glacier, and embracing **lodgement till** (plastered on the bed) and **melt-out till** (released from slow-moving or stagnant ice); deformation till (or glaciotectonite) resulting from shear deformation affected subglacial sediment and rock; **supraglacial melt-out till**, let down onto the substrate from the glacier surface. Because of the difficulty of identifying processes in a deposit, especially when they occur in tandem, some glacial geologists argue that the above terms are not helpful, and that basal tills should be grouped under the collective term **subglacial traction till**. This reflects current understanding that the **subglacial till mosaic** represents a range of interacting processes and sediments occurring at the glacier bed.

Till delta

A wedge of **till** and other **glacigenic sediment** built outwards in conveyor-like fashion from the **grounding-line** of an **ice shelf** or **ice stream**. Till deltas are usually recognised in offshore seismic data. The non-genetic term **diamict(on/ite) apron** may be preferred (*cf.* **trough-mouth fan**).

Till fabric (see fabric (till))

Tillite

The hard, rock (or lithified) equivalent of **till** (Figure T.14). Traditionally, the term has embraced a wider range of **glacigenic sediment** than the precisely defined definition of **till** and has included **glaciomarine sediment**. A more restrictive matching definition is preferred where tillite is solely the product of direct deposition from glacier ice. A related, but near-obsolete, term is **tilloid**, meaning a till-like rock of uncertain origin. Because of their relative scarcity in the geological record, tillites have been used widely for correlation and reconstructing former continents. The most widespread tillites are of late Precambrian (Neoproterozoic; c. 750 to 600 Ma) age and have given rise to the **Snowball Earth** theory of global glaciation.

FIGURE T.14 Tillite of Neoproterozoic age in the Mineral Fork Formation of the Wasatch Mountains, Utah. Note the poorly sorted nature of this rock and its resemblance to basal till (Figure B.10).

Till plain (see **ground moraine**)

Toe (term used mainly in North America; see **snout**)

Tongue

The lower part of a valley glacier or outlet glacier that extends below the equilibrium line (Figure T.15). See also **glacier tongue**, which applies to unconstrained elongate extensions of ice streams.

FIGURE T.15 Aerial view of part of the tongue of Unteraargletscher extending from the 4277 m peak of Finsteraarhorn, Switzerland. Prominent lateral and medial moraines and ogives are visible (2004).

Topographic inversion

The process whereby a landscape surface has reversed its elevation in relation to other surfaces. The process is common on a **debris-mantled glacier** and in **dead ice** areas (Figure T.16). A debris mound develops by differential ablation, the material slides off the mound into hollows, and the thickened debris in the hollow slows down ablation whilst ablation in the mound accelerates. The end result is that the hollow becomes the mound, and *vice versa*. The process is repeated at different rates over the whole debris-covered area of the glacier.

FIGURE T.16 Topographic inversion taking place on Lhotse Shar Glacier, Khumbu Himal, Nepal. Debris is sliding off the mounds into the ponds. Eventually the depressions will accumulate enough sediment to retard ablation relative to the surroundings (May 2003).

Tor

A remnant of an ancient land surface that has been subjected to frost-shattering or **periglacial** processes (Figure T.17). In **glaciated** areas, tors are believed to have two possible origins: (a) they formed on land by frost action exposed above an **ice sheet** as **nunataks**, for example, during the **Last Glacial Maximum**; (b) they survived from an earlier age and were protected or only slightly modified by later ice sheets which covered them. Accurate dating is crucial to correct interpretation of these features; the relatively new technique of cosmogenic nuclide dating aids in doing this.

FIGURE T.17 Frost-shattered, strongly cleaved volcanic rocks of Ordovician age comprise the tor of Castell-y-Gwynt (meaning 'Castle of the Winds'), on the Glyderau range, Snowdonia National Park, Wales.

Total strain (see cumulative strain)

Transection glaciers

Glaciers occupying a deeply dissected mountain landscape, where **valley glaciers** are interconnected. Individual glaciers flow from a **highland icefield** into a system of radiating valleys and spill over existing drainage divides. The overall pattern is web-like, and ice **diffluence** occurs where ice flow divides into two or more separate valleys (Figure T.18).

FIGURE T.18 Transection glaciers in Knut Rasmussen Land, south of Scoresby Sund, on the eastern coast of Greenland. In the background is a highland icefield and the glacier flowing into the foreground is Sortebræ, which is one of several interconnected valley glaciers in the region (July 2012).

Transfluence

The large-scale breaching of a mountain range by **outlet glaciers** emanating from an ice sheet, with all low points (**cols**) being occupied by the discharging ice (Figure T.19).

FIGURE T.19 The best example of transfluence today features the breaching of the Transantarctic Mountains by the East Antarctic Ice Sheet. This is Liv Glacier, one of the many huge outlet glaciers that breach the mountains and flow down into the Ross Ice Shelf, just visible at top left (December 1995).

T

Transient snowline

The snowline on a glacier at any instant. The term applies during the ablation season when the snowline is receding up-glacier (Figure T.20).

FIGURE T.20 Transient snowline on Ewigschneefeld on upper Grosser Aletschgletscher, Switzerland (August 2004).

Transverse crevasses

Crevasses that are orientated at right angles to the flow direction of a glacier. Commonly, they begin with a concave down-glacier geometry, before becoming straight (Figure T.21). Transverse crevasses are formed where there is a bedrock step and may be associated with an icefall.

FIGURE T.21 Transverse crevasses on Glacier d'Argentière, Mont Blanc Massif, France. These flow from right to left, with crevasses beginning with a concave down-glacier geometry, then straightening out, and then degenerating into an icefall at left (August 2008).

Transverse ridge (see **De Geer moraines**)

Transverse foliation (see **arcuate foliation**)

Tributary glacier

A glacier flowing from a side valley into a larger valley glacier (Figure T.22).

FIGURE T.22 An unnamed tributary glacier flowing into Steele Glacier, Icefield Ranges, Yukon (July 2006).

Trimline

A sharp line on a hillside marking the boundary between well-vegetated or well-weathered terrain that has remained ice-free for a considerable time, and poorly vegetated freshly exposed terrain. In many areas the most prominent trimlines date from the Little Ice Age or Neoglacial period (Figure T.23). Trimlines, in combination with moraines, can be used to reconstruct former glacier extent and thickness.

FIGURE   T.23 Prominent trimline above rapidly receding Kongsbreen (west), northwestern Spitsbergen, Svalbard. The pale coloured rock has been progressively exposed since the Neoglacial time (around 1900 AD). The dark rock above is highly weathered and carries lichen and scattered vegetation (July 2013).

T

Trough-head (synonym: trough-end)

The steep face quarried by a **valley glacier** at the head of a **glacial trough** (Figure T.24a and b).

(a)

(b)

FIGURE T.24 (a) The debris-mantled Glacier du Miage in the Mont Blanc Massif, Italy, occupies a long straight trough, at the far end of which is a trough-head (July 2007). (b) A trough-head at the head of the Buttermere Valley, English Lake District, with the peaks of Fleetwith Pike on the left and Haystacks on the right. A plateau icefield fed the main valley *via* the head of this trough.

Trough-mouth fan

A large-scale arcuate accumulation of sediment built out from the edge of the continental shelf when a glacier reached this position and deposited its load at the break of slope (Figure T.25). Typically these fans are tens of kilometres across.

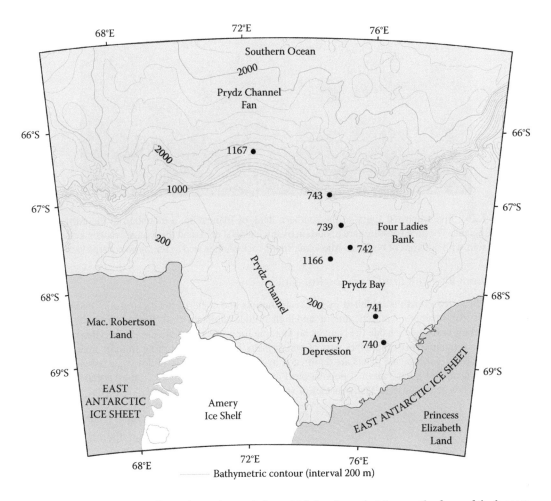

FIGURE T.25 The Prydz Channel trough-mouth fan, which has formed at the mouth of one of the longest-lived outlet glaciers in Antarctica, with a glacial history spanning 34 million years. This fan is well studied; several sites have been drilled by the Ocean Drilling Program (indicated by numbers) and extensive geophysical surveys have been undertaken. (Adapted from O'Brien, P.E. et al., 2004, Prydz Channel Fan and the history of extreme ice advances in Prydz Bay, Ocean Drilling Program, Leg 188.)

Truncated spur

A steep spur on the side of a **glacial trough** that was formed when a **valley glacier** widened and steepened the sides of the valley. The uppermost parts of the spur, which are not truncated, represent the remnants of a former river-valley system of interlocking spurs. Several spurs may occur together in a straightened valley (Figure T.26).

FIGURE T.26 Five truncated spurs on the southern flank of Blencathra, a mountain in the northern Lake District, England. The green pastures are underlain by thick glacial drift, and the orange bracken-covered areas have thin soils with outcrops of bedrock. Above the spurs, rocky arêtes lead to the snow-covered summit ridge.

Tundra (from the Sami *tūndâr via* the Russian)

A biome where tree growth is inhibited by low temperatures and a short growing season. Tundra occurs at high latitude (Arctic tundra, Antarctic tundra) and high altitude (alpine tundra). Tundra surrounds glacierised areas and is best developed outside the Neoglacial moraine limit. Arctic tundra usually overlies permanently frozen ground (**permafrost**), the upper part of which melts in summer, resulting in large areas of marshy ground. Biodiversity is low, and plants are dominated by shrubs and a variety of flowers, grasses, mosses and lichens, which support a low mammal population (Figure T.27). Alpine tundra is often not associated with permafrost and is better drained. Many species of vegetation are common to both.

FIGURE T.27 Arctic tundra on Ossian Sarsfjellet, northwestern Spitsbergen, Svalbard. On this ground, outside the Neoglacial limits, a Svalbard reindeer (*Rangifer tarandus platyrhynchus*) is grazing on characteristic high-Arctic flora. The glacier Kongsbreen is in the background (July 2009).

Tunnel valley (synonym: Rinnentaler)

A large subglacial, steep-sided channel cut into soft sediment or bedrock by meltwater beneath an ice sheet (Figure T.28). Tunnel valleys have flat bottoms and steep sides, and may exceed 100 km in length or 4 km in width. Tunnel valleys may occur singly, or as an anastomosing or dendritic network extending over large areas. Tunnel valleys represent larger versions of **Nye channels**, with which they share common characteristics such as undulatory long profiles, and overdeepened basins, denoting the impact of glacial meltwater under high pressure.

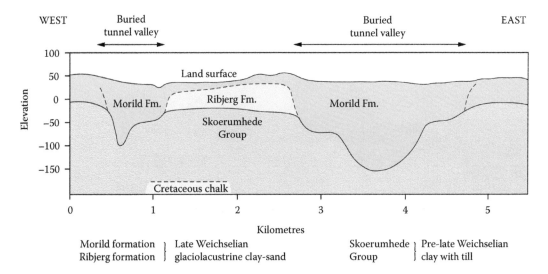

FIGURE T.28 Cross-section through tunnel valleys in the Verdsyssel region of Denmark. These tunnel valleys were formed under the outer part of the Weichselian Ice Sheet between 20 and 18 ka, and are incised into earlier Weichselian sediments. (Based on Figure 9 in Sanderson, P.B.E. et al., 2009. Rapid tunnel-valley formation beneath the receding Late Weichselian ice sheet in Vendsyssel, Denmark. Boreas 10.1111/j.1502-3885.2009.00105x.)

Turbidity flow/turbidite

A high-density, sediment-bearing current that flows down an underwater slope. The resulting deposit is a **turbidite**, characterised by graded bedding. In glacial environments the sediment is carried by subglacial streams, or within debris flows derived from subglacial till, before being released into a body of water (Figure T.29). Subaqueous sediment collapse on steep slopes can also generate a turbidity flow.

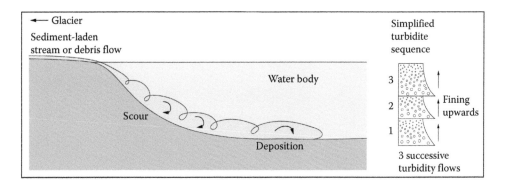

FIGURE T.29 A simple diagram to show a turbidity flow formed by a sediment-laden stream or debris flow entering a water body. The sedimentary products, turbidites, are a series of graded beds, illustrated on the right, although in practice they are commonly more complex than this.

Tuya (also known as Stapi)

A subglacially erupted volcano characterised by steep flanks and a flat or slightly domed summit (Figure T.30). Named after a volcano in Iceland, no longer enveloped in ice.

FIGURE T.30 Herðubreið, a mountain in northern Iceland, is a tuya formed beneath the ice sheet that covered most of Iceland at the Last Glacial Maximum.

T

U

U-shaped valley

The term generally applied to describing the cross-sectional profile of a **glacial trough** with near-vertical sides and a relatively flat bottom (Figure U.1a and b). In reality, true U-shaped profiles are rare, and open valleys with parabolic cross-sections are more typical of glacially eroded landscapes (see **parabolic form**).

(a)

(b)

FIGURE U.1 Two well-known glacial troughs with true U-shaped profiles. (a) The Yosemite Valley in California; the granite rock wall of El Capitan is on the left. (b) Lauterbrunnental as seen from the village of Lauterbrunnen, Kanton Bern, Switzerland. The vertical valley sides give rise to several impressive waterfalls, such as Staubbachfall on the right.

Unconformity

A discontinuity in the annual layering in **firn** or ice, resulting from a period when **ablation** cut across successive layers (Figure U.2). The term is borrowed from geology.

FIGURE U.2 A skidoo party passes beneath an ice cliff showing a trough-like unconformity in Granite Harbour, western Ross Sea, Antarctica (December 1997).

Upwelling (ice)

A flow of water under high pressure emerging at the surface of a **glacier** from an **englacial** or **subglacial conduit** (Figure U.3).

FIGURE U.3 Upwelling on the surface of a polythermal glacier, Austre Brøggerbreen, northeastern Spitsbergen, Svalbard. This water has evidently been in contact with the bed on the far side of the glacier where the bedrock is red sandstone (August 2009).

Upwelling (sediment)

A flow of water under high pressure from ice buried by **glaciofluvial sediment** (Figure U.4). Most commonly associated with **polythermal glaciers**, it is likely that sediment buries remnants of **dead ice**, which maintain a hydrological connection with the active glacier.

FIGURE U.4   Upwelling in glaciofluvial sediment near the snout of Midtre Lovénbreen, northeastern Spitsbergen, Svalbard. The supercooled water is here refreezing as it emerges, forming a mound of ice (August 1999).

Upwelling mound

An irregular mound of sediment created by an upwelling of the type above, typically on a **braid plain** in front of a glacier (Figure U.5). These rare phenomena indicate a hydrological connection with the **glacier** and comprise well-sorted gravel and sand.

U

FIGURE U.5   An upwelling mound about 3–4 m high surrounded by Aufeis on the braid plain in front of Fountain Glacier, Bylot Island, Canadian Arctic. The site of the water egress was in the middle of the mound and formed a radial pattern of pebbles and cobbles. The upwelling was no longer active in this location when the photograph was taken in July 2014.

V

V-shaped valley

Valleys that are V-shaped in cross-section are generally regarded as typical of river valleys with interlocking spurs. However, glacial valleys can also be V-shaped if the long profile is steeply graded. Most of the erosion is undertaken by subglacial meltwater. The distinguishing feature of a V-shaped glacial valley is its straightness, and there may also be remnants of **moraines** along its flanks (Figure V.1).

FIGURE V.1   A straight V-shaped glacial valley with lateral moraines holding back scree slopes, Burnett Mountains, Aoraki/Mt Cook National Park, South Island, New Zealand.

V

Valley glacier

A glacier bounded by the walls of a valley, and descending from high mountains, from an ice cap on a plateau, or from an ice sheet (*cf.* **alpine glacier**; **outlet glacier**) (Figure V.2).

FIGURE V.2   A valley glacier with prominent medial moraines, South Arm of Kaskawulsh Glacier, Icefield Ranges, Yukon (July 2006).

Valley train

A flat tract of glacial outwash material (gravel and sand) flanked by mountains (Figure V.3). The river is **braided** with seasonally high bedload, and unstable banks result in continual and channel switching.

FIGURE V.3   Convergence of two glacial rivers below Steele Glacier, Yukon, produces a valley train. The braid plain has both active and stable components, the latter occurring in the middle of the river.

Varves

Laminated sediment formed from fine-grained glacial debris entering a lake in suspension *via* a subglacial stream or a short glacial river. An individual varve is either a graded or a paired layer comprising sand+silt or silt+clay (Figure V.4). The coarser fraction represents summer sedimentation and the finer fraction settling out from suspension in winter. The annual nature of varves, if supported by an independent method of dating, allows thick accumulations to provide sedimentation rates and a chronology of environmental change.

FIGURE V.4 Thin section of varves in a core extracted from sediments in former glacial lake, Llyn Teifi, near Cardigan, south-west Wales. A pair of light (sandy) and dark (muddy) laminae forms a couplet representing one year's accumulation of sediment. The laminae are each approximately 2–5 mm thick. A small syn-sedimentary fault is evident top left. Larger grains may be ice-rafted debris.

Vegetation colonisation

When a glacier recedes, the freshly exposed ground becomes suitable for colonisation by glaciers. Moss and lichen appear first, followed by grasses and flowering plants, and then by shrubs and trees (Figure V.5) (see also **glacier vegetation**).

V

FIGURE V.5 Vegetation colonisation in the proglacial area of Vadret da Morteratsch between 1940, when the glacier snout was located over this area (indicated by the marker post), and the summer of 2002 when the photograph was taken. Various shrubs and tress have taken hold on the stony ground, with larch becoming the dominant species.

Velocity

The speed of, for example, a **glacier**, usually measured in metres per year or centimetres per day. In glaciers, flow is faster in the middle than at the sides and the upper part than at the bed. This attribute is the result of **internal deformation**. See also **glacier flow**.

Verglas (from the French *verre* + *glace*)

A thin layer of solid ice formed by the freezing of meltwater, supercooled raindrops or condensed water vapour on rock surfaces. Since it is difficult to see, verglas can be treacherous to the unwary.

Viscosity

Newtonian viscosity refers to a linear relationship between strain rate and shear stress. For comparison with plasticity shown graphically, see Figure P.12.

Volcanic bomb

A piece of lava ejected by a volcano during an eruption, which falls to the ground as a solid lump of rock. Glaciers on volcanoes may show a scatter of volcanic bombs (Figure V.6), commonly in association with **tephra**.

V

FIGURE V.6   A 1 m diameter volcanic bomb (andesite) and tephra on the small glacier in Ruapehu's South Crater, North Island, New Zealand. The bomb and tephra are the products of a small eruption in October 2006 (April 2008).

W

Warm glacier (see **temperate glacier**)

Warm-based glacier (synonym: **wet-based glacier**)

Warm ice

Ice that is at melting point and makes up the bulk of a **temperate glacier**. The temperature may be slightly below 0°C at the base of a glacier where the ice is under high pressure.

Water equivalent

The vertical extent or mass of water that would be obtained by melting an equivalent thickness or volume of snow, firn or ice. A metre of water equivalent (1 mwe) is formally defined in **mass-balance** studies by dividing a particular mass per unit area by the density of water (where 1 mwe = 1000 kg m^{-2}/density).

Waterline notch

A small undercut cliff, the top of which marks the level where an ice-dammed or supraglacial lake reached before draining. A prominent overhang may form if ice exposed to air melts less than the ice below water (Figure W.1).

FIGURE W.1   Prominent waterline notch with icicles around the edge of a drained supraglacial lake on George VI Ice Shelf, southwestern Antarctic Peninsula (November 2012).

Water table

The level up to which a glacier is saturated with water. Applying principally to temperate glaciers, the water flows along ice crystal boundaries and into cavities such as **moulins** and **englacial channels** until it reaches a stable level. The water table fluctuates on daily and seasonal cycles.

Wave ogives (see **ogives**)

Wet-based glacier

A glacier that has ice at the pressure melting point its bed, with meltwater facilitating sliding. **Temperate** and parts of **polythermal glaciers** are wet-based. Although consisting predominantly of cold ice, extensive wet-based conditions exist also beneath the ice sheets of Greenland and Antarctica.

Wet snow zone (see **slush zone** and **facies (snow)**)

Whaleback

A smooth, scratched (**striated**), glacially eroded bedrock knoll several metres high, and resembling a whale in profile (Figure W.2). Also referred to as a **rock drumlin** (*cf.* **roche moutonnée**).

FIGURE W.2 Whalebacks of granite recently exposed by the recession of Rhonegletscher, Valais, Switzerland (September 2009).

Whiteout

The meteorological condition of diffuse light in which visibility is reduced to zero by snow, especially during a blizzard (Figure B.18), or when a snow surface and continuous white cloud merge. The horizon disappears, surface irregularities are lost, and a person may become disorientated. Whiteouts on a glacier are common, and travel may have to be suspended, as hazardous features such as crevasses and supraglacial streams become invisible if drifted over.

Wind accumulation

Accumulation on a glacier that is controlled by blowing snow and topography. Long linear ridge of snow orientated in the direction of the prevailing wind form on the lee side of obstacles (Figure W.3a and b). Wind-controlled accumulation is most pronounced on, or at the edge of, extensive **highland icefields**, **ice caps** and **ice sheets**.

(a)

(b)

FIGURE W.3   Wind-controlled accumulation in Antarctica. (a) Long tails of wind-driven snow in the accumulation area of a highland icefield, culminating in Mt Wade (background), near Shackleton Glacier, Transantarctic Mountains, Antarctica (January 1996). (b) Wind-blown snow is the sole accumulation source for these small glaciers on Minna Bluff, overlooking the McMurdo Ice Shelf, Antarctica. Uninhibited southerly katabatic winds carry the snow over this volcanic ridge and onto the surface of the ice shelf. Two figures on the ice shelf are almost invisible in drifting snow (December 2010).

Wind scoop/wind scour

Hollow around rock outcrop caused by wind turbulence and erosion, or enhanced ablation (Figure W.4a and b).

(a)

(b)

FIGURE W.4   (a) Wind scoop around a prominent rock outcrop on Hallstätter Gletscher, Grosser Dachstein, Austria, where ablation is enhanced by proximity to the dark rock (Summer 2009). (b) The scouring effect of wind and deposition of snow in the lee of nunataks is well illustrated in this image of Harding Icefield, Kenai Peninsula, Alaska (Summer 2011).

Winter advance

A dynamic glacier may undergo a small advance in the winter season, even when it is undergoing general recession. On land, a series of **annual push moraines** may indicate a succession of winter advances. In a water body, lake ice or sea ice may be subject to deformation, fracturing and buckling (Figure W.5), indicating that in winter forward movement is not compensated for by ablation (melting or calving).

FIGURE W.5   Thin slices of sea ice, thrust up by a slight winter advance of Mackay Glacier, Granite Harbour, Victoria Land Antarctica (December 1997).

W

X

No entries for X.

Y

Yield stress (see **stress**)

The critical stress at which a material is able to flow. See graphical representation in Figure P.12.

Younger Dryas Stadial

A late stage glacial period (c. 13,000 to 11,500 years BP) that affected many parts of Europe, eastern North America and Arctic Alaska. This period is based on pollen records from southern Scandinavia, dated by radiocarbon methods. Because of dating errors and problems of precise climatic correlation, new research suggests that a better approach is to use the oxygen isotope ice core record from Greenland, where Greenland Stadial 1 runs from 11,500 to 12,700 years BP. The Younger Dryas period is, however, associated with 'fresh-looking' moraines in various mountain areas, such as the UK, where the term 'Loch Lomond Stadial' is often used instead (Figure Y.1).

FIGURE Y.1 Hummocky moraine dating from the Loch Lomond Stadial (Younger Dryas) cold phase of around 11,500–13,000 years ago, when glaciers were re-established for the last time in Britain: Coire Ardair, Grampian Highlands, Scotland.

Y

Z

Zastrugi (see **Sastrugi**)

Zone of suspension

The zone where debris is carried within the body of a glacier, without the particles coming into contact with the bed. Debris is derived from rockfall and buried in the accumulation area, and is little modified during englacial flow (Figure Z.1).

Zone of traction

The zone at the base of a **wet-based glacier** (the ice–bed interface), where debris is transported and, with bedrock, modified by crushing and **abrasion**. The debris becomes poorly sorted with a wide range of clast shapes and sizes, with **basal debris** within the body of the ice and **basal till** (or **subglacial traction till**) at the base of the glacier (Figure Z.1).

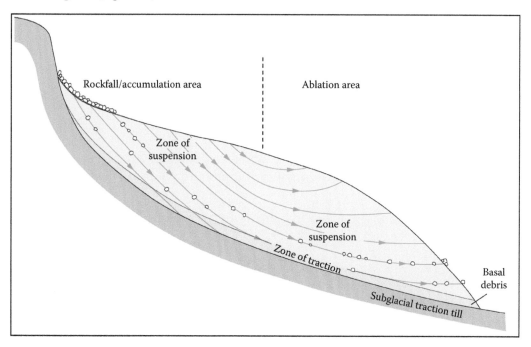

FIGURE Z.1 The zones of suspension and traction within a valley glacier, illustrated by a schematic longitudinal cross-section. Particle paths through the glacier are shown.

Z

Index

Note: **Bold** terms refers to geographical localities.

Printed and bound by CPI Group (UK) Ltd, Croydon, CR0 4YY

22/10/2024

01777614-0012